MONSTER FIRE AT MINONG

MONSTER FIRE AT MINONG

Wisconsin's Five Mile Tower Fire of 1977

BILL MATTHIAS

Wisconsin Historical Society Press

Published by the Wisconsin Historical Society Press
Publishers since 1855

wisconsinhistory.org

Front cover photo: Wisconsin DNR file photo by Bill Hoyt

Printed in the United States of America
Cover design by Andrew Brozyna, AJB Design
Interior design by Jim Primock

14 13 12 11 10 1 2 3 4 5

Library of Congress Cataloging-in-Publication Data

Matthias, Bill.
Monster fire at Minong : Wisconsin's Five Mile Tower Fire of 1977 / Bill Matthias.—1st ed.
p. cm.
Includes index.
ISBN 978-0-87020-447-0 (alk. paper)
1. Five Mile Tower Fire, Wisconsin, 1977. 2. Washburn County (Wis.)—History—20th century. 3. Douglas County (Wis.)—History—20th century. 4. Wildfires—Wisconsin—Prevention and control. I. Title.
SD421.32.W6M48 2010
363.3709775'15--dc22

2009023093

∞ The paper used in this publication meets the minimum requirements of the American National Standard for Information Sciences—Permanence of Paper for Printed Library Materials, ANSI Z39.48-1992.

Publication of this book was made possible in part by donations from

Wisconsin Department of Natural Resources

Wascott Volunteer Fire Department, in memory of Jamie Forbes

John Deere Construction & Forestry Co.

Gladys Matthias

John R. Liautaud

Ellen F. Buck

GR Manufacturing, Inc.

Biewer Wisconsin Sawmill, Inc.

Pierce Family Fund of the Minneapolis Foundation

L. Peter Patrick

Mary Beth Smith

Chester J. Hoversten and Clenora Hoversten

Richard A. Nelson

During research for Monster Fire at Minong, the author interviewed more than 135 people about their experiences on the Five Mile Tower Fire and its aftermath. He also incorporated his own recollections about the fire. Every attempt has been made to capture an accuate description of the events, participants, and experiences, and any inaccuracies or omissions are unintentional. If you are aware of any errors in this text, please send your comments to billmatthias@hotmail.com.

There are thousands of stories about this fire, and limited time and space to record them in this book. The author would like to hear from you if you had an experience in this fire that you would like to share for possible inclusion in a future edition of the book.

Dedicated to my parents, Vilas "Matt" Matthias (1914–2004), professor emeritus, University of Wisconsin College of Agricultural and Life Sciences, and Gladys (Wehrwein) Matthias.

Since they purchased property in the sandy soils of northwest Wisconsin in the mid-1950s, I have grown to love the pines, scrub oaks, and deep, cold, clear lakes of the region.

Contents

Introduction

I bend down to tee up the ball. On a warm day in 2006, my wife, Karen, and I are enjoying nine holes of golf at Fire Hill Golf Club in Gordon, Wisconsin. Chunky clouds race across the sky, pushed by spring's southwest-prevailing winds. Most people who play here don't know how the course got its name, but as my typical amateur opening drive slices into the jack pine grove, I know. I look across the expanse of green fairway to the forest in the distance and see the trees there, all the same size. I shiver as memories flood back.

Nearly thirty years earlier I was in this area under much different circumstances. The winds were blowing then, too, hot and full of smoke and soot. The trees were twice as tall, and our world was burning. Near 3:00 a.m. I was driving my pickup truck behind a group of Gordon fire trucks on Hill Road. As we turned into the driveway of the Russell Hill estate, the jack pines burned on both sides of the narrow path. We had to close our windows so our arms wouldn't burn. Suddenly we came to a clearing, where the main house stood ready to burst into flames. Two garages, a painting studio, and a barn were already on fire. The grass, pine needles, and trees in the compound were bright with flames. I jumped out of the truck to help haul fire hose. Only a few feet away, a fifty-foot-tall red pine exploded. Several of us scrambled away from the tornado of sparks and flying embers. The noise was so loud we had to shout to be heard. I looked up and saw that stately tree turn into a swirling, flaming beast, shooting a spiral of fire one hundred feet into the sky. The falling chunks of burning bark and twigs immediately set off hundreds of other fires in the dry leaves, needles, and brush. We ran from the deadly shower. Our small group had saved the Hill home, but all of the other buildings were reduced to flaming skeletons as we watched the roofs cave in.

That was in 1977. Along with hundreds of others, I had been fighting a monster fire for twelve hours, and it still wasn't out. I was dirty, smelly, tired, hungry, and thirsty. Wherever I looked across the fire scene, spread over twenty-two square miles, I saw death of animals, destruction of forestlands, flames here, blackness there. Trucks, men, women, equipment, dozers, and plows. People hunched over from the weight of heavy, water-filled backpack can sprayers. Men and women with shovels following plowed trails to put out lingering flames.

Two years earlier, I had been selected by the five-member Northwood School District Board of Education to be its district superintendent. Northwood was a small district; I was the administrator for two principals, forty teachers, and five hundred students in two schools with ten bus routes. Our district was spread over 430 square miles of forestlands, a few farms, and many beautiful lakes. Student density was just over one student per square mile. Everyone living in the region was attuned to rural life, enduring such realities as long travel distances, few close neighbors, severe cold in the winter, and swarms of mosquitoes, wood ticks, and biting deerflies come summer. But such remote living offered many benefits. There was hunting in beautiful wooded settings, fishing in clear, cold lakes, and the enjoyment of regularly seeing deer, bear, coyotes, foxes, eagles, and osprey. And as far as the eye could see, there were trees.

I also was a private pilot and often rented Piper airplanes at Bong Field in Superior and flew south to look over our region. I would marvel at the trees carpeting northwest Wisconsin from horizon to horizon—so dense it was hard to spot major roadways from the sky.

In the mid-1950s, my parents had purchased red pine plantations and oak forestlands in Wascott, between Pickerel and Bond Lakes. I had grown to love these lands and lakes, and when Karen, our young son, Mark, and I returned from a year in Tripoli, Libya, where I was the principal of the Oil Companies School, we dreamed of the possibility of living and working in the Wascott area. When I was hired as superintendent, we moved into an A-frame cabin we had built on Bond Lake six years earlier.

Having been a forestry student in my early days at the University of Wisconsin–Madison, and now living so close to the land—eating venison and fish, burning wood I cut and split myself, and wishing to help protect the precious resource of forestlands—I was acutely interested in firefighting and forest protection. I worked with the school board, staff, students, and Department of Natural Resources (DNR) rangers Bill Scott from Minong and Barry Stanek from Gordon to establish two student forest firefighting crews made up of mostly juniors and seniors at Northwood High School. The school board had approved the necessary policies, and parent permission slips were on file. Parents' fears were allayed by the knowledge that the youths were trained properly and covered by state insurance policies, including workers' compensation.

The students on the primary and secondary crews were more than ready to be called out of classes to help on a fire. When the school intercom called for them, the fire crew members flew through the classroom doors to their lockers to don work boots. Some were assigned to go to the kitchen and pick up the premade frozen sandwiches and grab a case of half-pints of milk from the cooler. The fire crew members would jump in a DNR van driven by a teacher and head to the Minong Ranger Station for tools, orders, and directions.

These dedicated firefighting students, some twenty of them, along with sixteen hundred other individuals, would spend sixteen hours fighting the Five Mile Tower Fire on April 30 and May 1, 1977. After the weekend conflagration, I canceled high school classes, and more than fifty of us spent the next three days doing the dirty, hot mop-up work. Students from Hayward, Spooner, and Solon Springs High Schools also worked on the fire and the mop-up operation. After laboring all day in the black ash, the students stumbled onto the school buses and rode home looking like coal miners. A large swath of land in our school district had burned, and residents from miles around had dropped what they were doing and risked their lives to help. It truly was citizenship in action.

Standing on the raised green surrounded by sand traps almost three decades later, I look to the south and recall what happened twenty-nine earlier as though it were yesterday. I remember the huge billowing cloud of smoke gathering in intensity every minute as the fire slammed northward before strong, gusty winds of seventeen to twenty-two miles per hour. I remember the hot, dry winds, almost like those of the Libyan desert. I didn't realize at the time that my life would change in the days and weeks that followed.

A few years after the Five Mile Tower Fire, I became one of the first members of the new Wascott Volunteer Fire Department. We started with little money, obtaining army surplus pickup trucks and lumbering army six-by-six trucks converted into tankers. When we built our first fire hall, I helped with the construction along with many others, crawling on the walls and roof rafters high above the ground. Thirty years later I am still on the Wascott Fire Department. Today we have modern equipment and radios and fifteen firefighters ready and trained to go on wildland or structure fires.

In the past half century in Wisconsin, the Five Mile Tower Fire was the largest fire started by a single ignition point—a match for a campfire intended to cook hot dogs. The fire destroyed almost fourteen thousand acres and burned sixty-three structures. It also caused a firestorm of changes in the way the DNR and volunteer fire departments fight fire, leading to significant improvements. I had a rare insider's view of the fire and its aftermath. I had worked on the fire for more than fifty hours; some of my DNR friends worked it for sixty hours without sleep. One month after the fire, I was the only non-DNR person invited to attend the DNR's "fire review and critique" session held at the Spooner Experiment Station meeting room. I heard the stories of the many rangers who fought that fire heroically and the dozer operators who risked their lives.

In the thirty years since, the fire has surfaced again and again in conversations I've had with citizens, volunteer firefighters, and DNR personnel. Clearly it had been a searing experience, both literally and in the memories of everyone who lived near Minong during those dark, dusty, dry, flaming days of 1977.

As the thirty-year anniversary of the fire approached, I knew that no one had researched and written the story of the Five Mile Tower Fire. The story needed to be told, and I decided to be the one to tell it—I owed that to myself and to all those who risked their lives fighting the flames. My research began in 2006 with a visit to the basement storage room of the Cumberland DNR building. There I opened a large, dusty cardboard box labeled in marker pen, FIVE MILE FIRE, and I began my journey. As I sifted through the lists of names, invoices, reports, copies of letters, and other documents and conducted my first interview, with recently retired DNR area ranger Ed Forrester, my beliefs were confirmed—this was a historic, record-setting fire and an important part of Wisconsin's history.

The Five Mile Tower Forest Fire started at 1:20 in the afternoon of Saturday, April 30, 1977, and burned out of control for sixteen hours. In the thirty-three years since, there has not been another wildland fire that large in Wisconsin. This is the story of that fire.

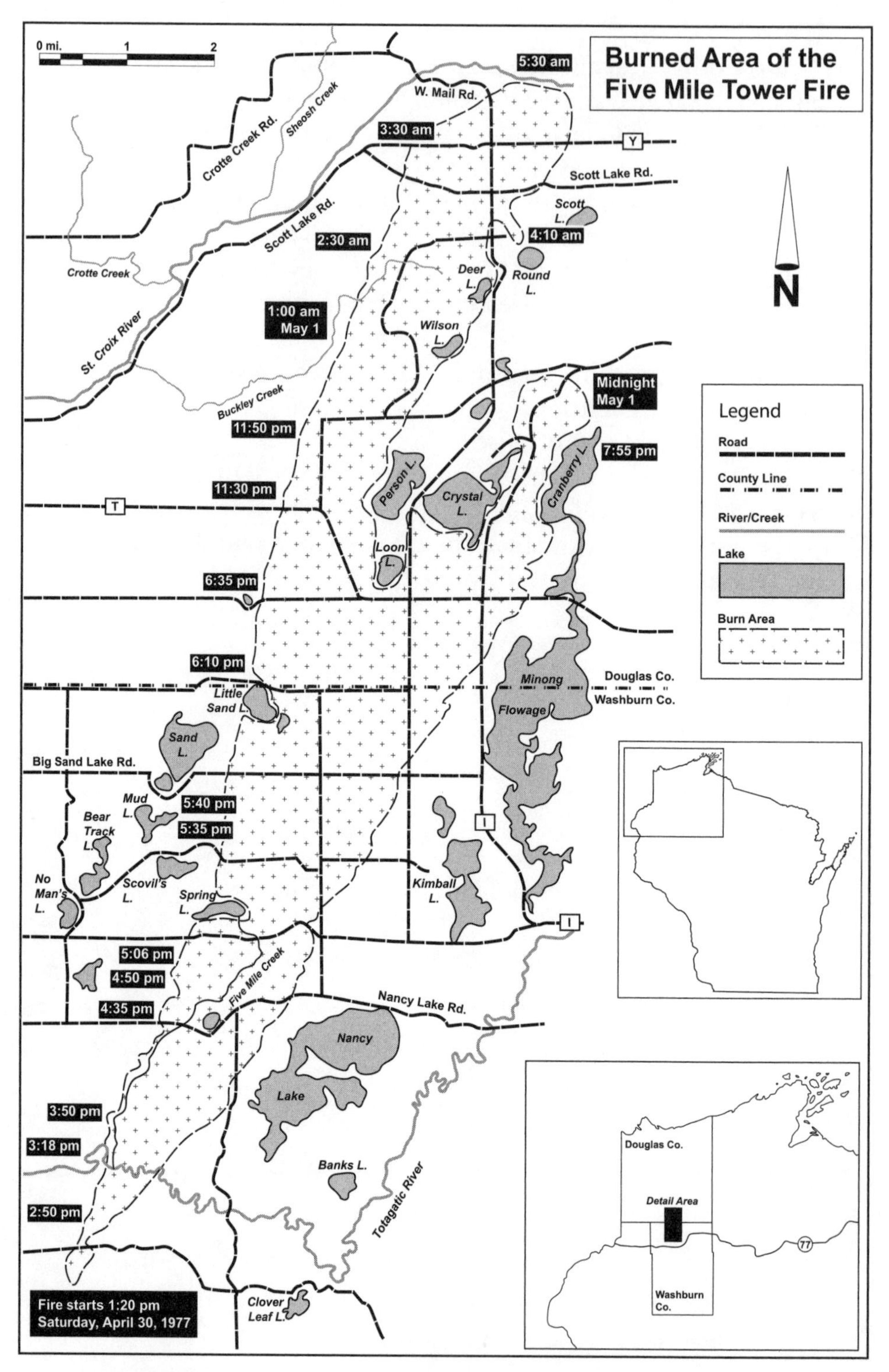

Map by Joel Heiman

Part 1

Anatomy of a Fire

The Family's Land

When the family arrived at their property on Friday, April 29, 1977, they had no way of knowing that this piece of land they treasured would be remembered for decades as the epicenter of one of Wisconsin's most important wildfires.

The 132 acres of prime deer hunting land was the pride of the Schultz family. John Schultz had just purchased the parcel in the town of Chicog in Washburn County, a few miles north of Highway 77. Now they had driven up from their home near Shell Lake to spend the weekend camping with their son and some friends. Nestled between the Namekagon and Totagatic Rivers, the acreage was excellent habitat for deer, bear, grouse, and many other creatures of the forest. In the distance, visible over the tops of the trees, the DNR's Five Mile Fire Tower stood sentinel. The jack pines and scrub oaks and land features made this an ideal property for the hunting enthusiast: bottomlands with rivers, swamps, high hills, and fallow farm fields covered in grass, with an occasional jack pine sticking out like a weed in a garden.

This land had changed greatly over the previous two hundred years. In the early 1800s, northern Wisconsin was mostly covered with huge white pine trees, mature and stately, hundreds of miles in every direction. By the 1840s, this expanse of towering trees with massive trunks three feet across was seen by lumber company owners not as hunting land but as means to making many millions of dollars. Within a few decades, most of these giants hit the ground to feed the insatiable appetite of a growing nation.

As the cutover era came to an end, much of northern Wisconsin looked like a plucked chicken. No large trees remained, and piles of slash containing treetops, branches, and dead stumps were everywhere, making the land look forlorn and worthless. Land speculators advertised cheap acreage in the region as good for farming. They advertised in Eastern and Western Europe and Scandinavia for farmers to start fresh here with their own land and be their own bosses. Many came, pulling and blasting the stumps and developing farm fields in the sandy soils and short growing season of the north. They built their barns and homes and fought hard for years to eke out a living. Some of the forest soils were rich for a time, and many settlers made it for a generation or more. But the soil was soon worn out, and the homesteaders sold out in frustration, moving to the nearby towns and cities for jobs.

On these former farm fields, new landowners established tree plantations in the 1940s and 1950s, making the poor soils of northern Wisconsin work again, growing jack pines and red pines for Christmas trees, pulp for paper, and bolts for lumber or telephone and electrical power poles. It was on one of these fields, covered in tinder dry grass and an occasional jack pine or scrub oak, that John Schultz made his campfire.

An Ominous Morning

Bill Scott rose from bed on Saturday, April 30, and looked out the window, a habit honed from years of work as a DNR forester/fire ranger. He saw the wind tossing socks left on the clothesline overnight. As he stepped outdoors, he noticed the early-morning breeze was from the southwest—unusual for this time of year. Humidity was low; there was no morning dew on the grass. Bill lived in the tiny village of Minong, population 550, with his wife, Kathy, and two sons.

Tall and stately, with blond hair, Scott was well known in the area. He had worked with the growing Minong Fire Department and its chief, Harold "Smokey" Smith. He was also an active parent in the Northwood School District. Scott was often called upon to manage study groups, chair special meetings, lead efforts for needed additional school facilities, and work on committees to accomplish local fire protection.

Since 1976, Scott had worked with me to establish high school crews made up of members of the junior and senior classes to fight forest fires. These students were trained in firefighting tactics and had been called to several small fires in the area. Some who showed the greatest interest were hired to work on weekends as standby line crew workers using backpack water cans weighing forty-five pounds when full. A few were picked to ride along with rangers as they made their patrols.

Bill Scott was in charge of the Minong Ranger Station and the men who operated its two heavy-unit tractor plows, trucks, and other equipment. During times of low fire danger, he and other rangers worked on forestland management and administration of private Forest Crop Law lands.

On this windy, dry morning, the Minong Ranger Station, nestled in the balsam trees next to State Highway 77 on the east edge of the village, was on full alert. Scott went to the station, got on the radio, and called for the weather report. It was bad. Humidity was at a desertlike reading of only 23 percent. There had been no rain for the past nine days. Then came killer news: southwest winds of seventeen to twenty miles per hour were expected. Scott checked his equipment and filled the DNR ranger-issue four-by-four heavy-duty pickup truck with fuel.

He greeted equipment operators Pete Paske, George Becherer, and Dale Remington as well as seventeen-year-old Northwood High School student firefighter Bud Schaefer when they arrived early that morning. Two other Northwood students, Tom Frye and Mike McShane, also checked in; they worked on weekends as standby firefighters. The three boys arrived with leather boots, jeans, and DNR-issued fire-retardant shirts. As they started their workday, they talked about the upcoming junior prom, which would be the next Saturday evening at school.

Paske and Becherer greased and oiled their yellow heavy units and filled them with diesel fuel. The men started and warmed up the large trucks, which pulled trailers, each carrying a dozer earth mover and furrow plow. Scott sent Becherer out in his older unit to Webb Lake, fifteen miles west of Minong in sand country dotted with lakes and pine forests, in case a fire should start in that region. Paske stayed at the ranger station with his new John Deere 450 heavy unit pulled by a three-ton truck. Remington was sent to climb the Lampson Fire Tower about eight miles to the south and watch for smoke. The high school boys readied the van and their hand tools, filled the backpack sprayer cans, and otherwise busied themselves as they listened to the high-band radio transmissions.

Bill Scott was thinking about his role that morning and the many hours he had spent during winter sessions in Tomahawk at the DNR fire-training headquarters. If a fire in his area got out of control, he would be the fire boss until the area ranger showed up; then he would probably be in charge of directing equipment and firefighters on the moving head, or front edge, of the fire. Other, more senior ranger administrators would be assigned to stay at fire headquarters and be line boss, plans boss, or service boss. Scott could visualize the charts and organizational flow diagrams showing exactly how the overhead team would organize as such a fire progressed.

DNR rangers know well that the goal is to push a fire's shape into an elongated balloon (or as some rangers describe it, the shape of an otter, long and narrow) by constantly pinching the sides of the burning area with dozers and plows. Experience had taught them that a crown fire, which leaps from tree to tree high in the air, usually cannot be stopped at the head because it is just too fast and too hot. Should a fire start, as other rangers arrived on scene with their heavy units, they would be assigned to divisions along the right or left flanks of the fire. Everyone would move with the flames, always trying to pinch the fire, narrow it, and finally stop it.

John DeLaMater was excited and worried. He rose just after sunrise Saturday and looked out the window, noticing the young leaves on trees in his yard trembling in unusual dawn winds. He was concerned about the conditions. But this was also a big day for DeLaMater, a friendly faced man with dark, wavy hair. His wife, Glenda, and newborn daughter, Jodi, were at the new Hayward Hospital, waiting to be picked up and brought home. Jodi had been the first baby born in the new facility, a bit of a surprise for the nurses, who had to scramble in unfamiliar cabinets for delivery instruments.

DeLaMater called the hospital and received permission from the head maternity nurse to take his wife and child home. An hour later, John and Glenda were getting Jodi settled in her new surroundings and enjoying a few moments of peace and quiet.

Then the phone rang.

DeLaMater was already thinking of changing into his steel-toed leather boots and fire-retardant shirt. He wondered what had happened this early in the forestlands of northwest Wisconsin. But it was the nurse calling. She was apologetic, saying the three of them had to return to the hospital because baby Jodi needed injections that the charge nurse had forgotten.

On this second trip to the hospital and back, DeLaMater was increasingly agitated and distant, thinking of his awesome responsibilities as the Hayward Area DNR fire ranger in charge of protecting thousands of acres of timber, swamps, plantations, and woodlots. This also made him among those responsible for protecting the lives of thousands of human beings; their homes, cabins, and outbuildings; and the many forest animals.

As he steered into his driveway, he saw a DNR pickup truck parked on the street, engine running. Sawyer County forester Chuck Adams was in the driver's seat. Glenda looked at her husband and knowingly kissed him good-bye. She knew the drill. It was dry, and the wind was blowing. Her husband was in charge of eight DNR ranger stations stretching east to west from Hayward to Grantsburg and north to south from Ladysmith to Minong—a huge swath of northwest Wisconsin, sixty by eighty miles. He was on watch over 3,072,000 acres of Wisconsin's precious forestlands. He would have to travel forty miles to the jack pines and sandy soils, where the fire threat was the highest. Glenda didn't know it then, but she and the baby would not see John again for five days. Giant fires burned in Wisconsin in 1871, 1884, 1891, and 1959, but on that first day home for baby Jodi, John DeLaMater would become the fire boss for the largest single-source fire in 106 years of Wisconsin history.

Driving out of Hayward, DeLaMater reviewed the orders he had made at 5:00 p.m. the day before and wondered if he had forgotten anything. Adams was listening. All national fire-danger factors were in the "high" range. DeLaMater had ordered all fire towers to be staffed at 10:00 a.m. on April 30 and requested a DNR airplane to be flying patrol in his region by 1:00 p.m. He had called the dispatch center and ordered all trained DNR personnel assigned to the "overhead" team to be on standby. All ranger stations were to have extra crews ready to go. John had already requested all county foresters and their John Deere 450 plow units as well as crews from the U.S. Department of Agriculture Forest Service and their dozers stationed at Hayward to be on standby.

The two rangers headed east on State Highway 77 toward Minong. The two-way radio was quiet. Dust in the air was reducing visibility to less than ten miles on the ground. The wind was picking up.

Barry Stanek, the Gordon DNR ranger, had the station everyone wanted—everyone, that is, who was not afraid of fire and who

An old-style dozer similar to many of those used on the Five Mile Tower Fire pulls a steel-wheeled plow through the trees.

Wisconsin DNR file photo

wanted to be smack-dab in the middle of sandy soils of Norway pine plantations intermixed with jack pines and scrub oak groves. Stanek had the responsibility of protecting not only the forestlands but also valuable shoreline property—hundreds of lakes ringed with cabins and homes. Gordon had its own fire tower high on the hill just west of the hamlet boasting the same name and two hundred residents. State Highway 53, the main north-south route between Eau Claire and Superior, bisected the town.

The Gordon Ranger Station was surrounded by large pines and firs on the southern shoreline of the Gordon Flowage, also called the St. Croix Flowage, a large body of slow-moving water thirteen miles long and a mile wide.

Stanek had just taken over for Major Mellor, who had retired as the longtime ranger at Gordon. The new ranger was a twenty-six-year-old forestry program graduate and military veteran, on the job for only four months. The previous year, Stanek had worked with Mellor on many fires, as Wisconsin had suffered from an extreme drought starting in 1976.

The conditions were in the "extreme" threat category. The drought was so severe that woodland soils and swamps had dried up, offering no relief from small blazes in August, September, and October of 1976. The fires had "duffed" into the ground, burning into the soil, peat, and roots, deep in the earth. Duff, the layer of organic material on top of the ground, was powder dry going into the winter, and Stanek and Mellor had been called to several fires even after it had snowed. Farmers trying to burn stumps, brush, and trees from fields would call the Gordon station saying their fires would not go out. Snow cover was less than normal, and the fires kept burning down into the peat.

Stanek made people feel confident that he knew his business and was serious about protecting natural resources. He was an excellent public speaker. When lake association groups and other organizations called upon him to give talks and slide shows about DNR activities, he never missed the chance to emphasize the importance of citizens making their property less vulnerable to fire. He taught techniques such as creating defensible space around cabins by removing vegetation, and widening driveways and providing turnarounds for emergency equipment and trucks in case of fire. Stanek was a small-town hero of sorts, one of the leaders on the Gordon Volunteer Fire Department and a highly trained and experienced emergency medical technician. His wife, Lorraine, and two young sons were active in their church and community.

On this Saturday morning, Stanek was at the ranger station early, checking equipment, tuning his radio, and noticing that the day was quiet, at least so far. His operators, John Kiel and Bobby Hoyt, drove aging Cletrac tractors hooked to "sulky" plows with steel wheels. The Gordon station still did not have the new John Deere 450 heavy unit. Each small, steel-tracked dozer had a rollover canopy and a small water tank with hose reel for dousing small fires. Blaine Peterson, a student at Northwood High School and a member of its primary fire crew, was also at the station that day. He served as Stanek's assistant, riding along in the ranger's four-by-four pickup. Ron Kofal, another heavy-unit operator, was just coming back to work after several weeks off for knee surgery.

Stanek had seen plenty of fires, studied fire weather data, and received extensive training at the Tomahawk center. He knew this day was a bad one. Spring 1977 was as dry as any time in recorded Wisconsin history. He thought back to the previous winter and the many nights he spent training the local volunteer fire departments, predicting that a big fire could occur at any time—training focused on how volunteer departments would cooperate by helping to protect buildings while the DNR concentrated on forestlands.

Now, the Gordon Ranger Station was ready. The fire lookout person climbed the steep steps to the tower enclosure, opened the windows, and made sure the DNR telephone was working. The station was teeming with life as the wind increased and the sun heated the atmosphere. Soon the station would be empty, all staff called to assist on a fire developing near the confluence of State Highway 77 and the Namekagon and Totagatic Rivers, fifteen miles away. The crew would not return to home base for three days.

The Match and the Horror

John Schultz and his camping party were hungry. The group had been busy all morning cutting branches, hauling brush, and hiking Schultz's newly purchased property. Hot dogs were on the menu for lunch, and at about 1:00 p.m. Saturday Schultz carefully placed rocks in a three-foot ring and dug six inches down into the grassy earth, once part of a farm field. The jack pines and oaks were at least one hundred yards away, so this seemed like a safe place to start the fire. Everyone gathered around with hot dogs skewered on whittled sticks, waiting as Schultz knelt down and struck a match. It was the worst decision of his life.

The little campfire started, and a gust of wind blew a burning leaf outside the ring. Immediately, the long, tinder-dry grass caught fire, pushed by seventeen-mile-per-hour winds from the southwest. It happened so fast they could hardly believe it.

Remnants of the rock-ringed campfire that ignited the monster fire.

Wisconsin DNR file photo

They frantically fought the fire with coats, jackets, shirts, shovels, and what little water they had on hand. The cabin Schultz and his wife planned to build on the land someday was just drawings on paper, so they had no well or water hose. The fire danced across the field, toward the wood line. There was screaming and yelling, gasping and grunting as the small group fought the fire, but it was moving steadily in the brown grass, already leaving black ash behind. Soon their clothes and skin were sooty from the work. The fire was galloping like a wild stallion. First it was here, then it was there. When someone hit one spot with a shovel, it caused five more little fires to start as burning grass flew in all directions. They tried to rake the flames, and it got worse. Their muscles ached from their futile efforts.

The Schultz party felt a mix of horror and fear. Several cried tears of frustration and shame. The blaze started by their single match was out of control. John rushed to help his wife and ten-year-old son into their car. The flames were headed to the trees near the driveway, and they had to evacuate. His friends left, too, driving in fear through the smoke and rising flames, trying to get out before the jack pines caught fire and endangered their lives.

His wife begged him to come along, but John would not leave. He felt remorse and guilt and a sinking feeling that his life had changed dramatically the instant he lit the match. He had to help stop this blaze. He told his wife to drive fast to Pappy's Tavern on Highway 77 to call the Minong Fire Department. His friends were already on the CB radio, calling the nearby Shangri-La Bar, six miles away. The resulting call from the Shangri-La to the Spooner DNR dispatch center was labeled Number 131 (there had been 130 fires prior to this one in 1977).

After getting his family and friends out, John ran back to the fire and grabbed a rake. He continued his futile efforts to halt some of the flaming front that now looked like a marching red army. Flames had almost reached the dry jack pines on the edge of the field. Then he heard a truck coming and ran toward it, waving and yelling.

It was Minong ranger Bill Scott and his high school assistant Bud Schaefer in the DNR truck. They had heard the call from the Five Mile Fire Tower just three miles away, where observer Cleo Cox had seen the smoke. Scott was first on the scene, and rangers John DeLaMater and Chuck Adams soon arrived. Scott told Schultz to give his name and address to the pair, and then he rushed to turn around in the tight driveway, pushing over two small jack pines with his rear bumper. He had to get out of there before the truck burned up. When the fire reached the young jack pine grove, the flames would be magnified a hundred times in both temperature and size.

John Schultz ran down the sandy road, away from the flames, and soon came upon DeLaMater and Adams in the Hayward DNR pickup. Adams jumped out and took down Schultz's identification information, and then he raced off to establish a fire headquarters and begin building an overhead team to manage operations.

The black column of burning pine smoke billowed high, and folks could see it from twenty miles away. Citizens from Minong, Chicog, Webster, Gordon, Wascott, Hayward, Spooner, and Solon Springs quickly started showing up to help. Within the first hour, two hundred people were there to assist, including John Schultz, working on foot with backpack cans and shovels, trying to keep the fires from jumping the dirt furrows made by the plows. Schultz would work like a dog all night and into the light of Sunday before turning himself in—just one of the workers that eventually numbered more than sixteen hundred, doing everything he could to halt the fire he knew too well. He loved this country and felt horrible regret for the havoc the fire was wreaking. He wasn't going to hide, so he walked up to someone wearing a DNR shirt and said, "I am John Schultz, and I was the one who started this fire."

Warden Milt Dieckman was guarding an intersection at Highway T on Sunday morning when a dirty, blackened, tired volunteer firefighter walked up to his car and identified himself as the one who had started the fire. Dieckman reported that John Schultz was humble and felt terrible about starting the fire. As Schultz told the story of how he and his family had tried to put out the flames, Dieckman took down the information; he would give the statement to his warden supervisor, Joe Davidowski, the next day.

On the Ground, In the Air

Barb Raasch, the Cumberland area fire dispatcher, had arrived at her cubbyhole dispatch office in the old Spooner Ranger Station at 9:00 a.m. that Saturday and settled into her chair. Her door was always open, because part of her job was to help folks who dropped in to the ranger station for burning permits, hunting licenses, and Smokey Bear literature.

On one side of the room the state telephone system switchboard with about twenty phone jacks and a spaghetti pile of cords and plugs would allow her to contact ten fire towers and DNR ranger offices around expansive northwest Wisconsin. She also had a regular telephone on her desk, next to the two-way radio.

Raasch was wired to the second-largest phone system in Wisconsin in 1977. The DNR had thick iron wires strung on hundreds of thousands of twenty-foot poles, connecting almost all of the fifty-two ranger stations with twelve regional dispatch centers all over the state as well as all one hundred fire towers. These heavy-duty phone wires crisscrossed the state for the purpose of forest fire communication and control. The equipment was mostly government surplus: mechanical switching devices and 1934 Graybar ringers, much like the old wooden crank phones used in America during the first half of the twentieth century. The system was labor-intensive, and each DNR station had to do its part to keep it maintained by fixing snapped wires, brushing rights-of-way, and chainsawing trees that fell over the lines. Lightning strikes were common. The college-educated DNR forester-rangers also had to be telephone repairmen to keep the system operating.

Filling the rest of Raasch's space were wall maps, a desk with a typewriter, and her DNR radio set. When rangers radioed Raasch, she would use either the state telephone lines or the public telephone to relay the information from the ranger in the field to local fire departments, police, or other, more distant ranger stations. The DNR forest rangers had radios in their trucks, but they did not have the frequencies to call anyone but other rangers or Raasch back at dispatch. They had no telephones in their trucks.

Raasch's position required that she live within ten minutes or ten miles of the Spooner Ranger Station. During dry seasons, if a ranger was called out after business hours to investigate reported smoke or fire, Barb was summoned to report immediately to her dispatch center. If the ranger needed local volunteer fire department or law-enforcement support, he had to radio Spooner dispatch. She was the rangers' crucial connection to the outside world.

When she was hired in 1976, Raasch was the first woman fire dispatcher in the Wisconsin DNR, and at least one individual told her she would have to work twice as hard as the others to prove that a woman could handle the phone lines and radio system.

That morning at 10:00, she had greeted each of the ten tower observers by the DNR's telephone lines and recorded the daily weather reports. In every case, the data showed it was a day to watch. When the first call came in from the McKenzie Tower, followed immediately by one from the Five Mile Tower, Raasch obtained the compass headings, went to her map on the wall, and connected the string. The fire was in Section 6 of the town of Chicog, Washburn County.

Sheriff's departments in three counties quickly sent out squad cars. Ambulances and volunteers were notified to report to a fire headquarters being assembled on the southwest corner of Highway 77 and Thompson Bridge Road, west of Minong. All conservation wardens in northwest Wisconsin were called to action. DeLaMater was issuing orders every few minutes by two-way radio. Dozens of entries were logged in the dispatch summary during the first hours of the fire. Raasch was a busy woman with awesome responsibilities.

She would work for the next sixteen hours, dispatching hundreds of requests from the fire boss and helping to orchestrate the largest buildup of firefighters, supplies, and equipment in Wisconsin fire history.

At 5:00 a.m. Sunday, after working nonstop for twenty hours, Barb Raasch would head home for some rest. But, too keyed up to sleep, she would return to work at 7:00 a.m. and stay for another ten hours.

At about the time Barb Raasch was making her initial contacts with tower observers on April 30, three DNR Beechcraft T-34 airplanes were cruising the skies of northwest Wisconsin. The DNR-employed pilots had been called into action by the area rangers to patrol for forest fires, to spot any smoke early enough for quick suppression.

Bud Erickson, Jim Dienstl, and Eugene Drinkwater were experienced former navy pilots. All three had been in war and had flown to hell and back many times for the navy and survived. On this hot, windy day they were challenged—but at least no one was shooting at them. They flew the state-owned planes, taking off from the Shell Lake Airport, where the state had two hangars.

Erickson had flown for the Wisconsin DNR for fourteen years. With the navy, he had flown off aircraft carriers in WWII in the South Pacific. He loved to talk about flying, and his enthusiasm was infectious.

Dienstl had flown A-4 attack bombers off aircraft carriers in Vietnam, surviving 150 missions over enemy territory. He was the only pilot in the Wisconsin DNR to have earned a master's degree in forestry. On the day of the Five Mile Tower Fire, he was flying patrol for the Cumberland region. Erickson and Drinkwater were flying far to the northeast for the Brule region and Park Falls region in order to cover the entire northwest part of Wisconsin.

The single-engine Beechcraft was sleek, flying at speeds approaching 130 miles per hour. As Dienstl flew over lakes, rivers, and forestlands of the Spooner and Minong area, he was thinking about the weather reports calling for high winds and low humidity. The updrafts created by rising air from the earth being heated by the sun caused the small plane to shudder at times from turbulence.

At 1:30 p.m., Dienstl's radio crackled to life. He was called to fly to an area west of Minong and north of Highway 77 where smoke had been reported by both the Five Mile Tower and the McKenzie

Three DNR spotter planes fly over Shell Lake in 1977.

Wisconsin DNR file photo

Tower. He banked his wings, dove to 1,000 feet, and soon could see the wispy column of smoke in the distance. The beginnings of this fire looked innocent from above, but as Dienstl was circling, he saw it hit a row of jack pines. The flames were propelled by the wind and seemed to blow up instantly, with glowing shards and clumps of burning embers thrown in all directions. He felt a lump in the pit of his stomach as he saw from the air what others could not—this was going to be a dangerous fire.

Back on the ground, Harold "Smokey" Smith was working as a heavy-equipment operator. Smokey was also Minong fire chief, and proud of it. He walked with confidence because he knew what he wanted and made sure everyone in the department understood his vision. He had been chief since 1972, and with five years of command experience, he was building the Minong Volunteer Fire Department into a formidable unit. By 1977, the department had two pumpers, one four-by-four army pickup truck, and two tankers for hauling water. All the trucks had pumps and hoses, so each could spray water.

As he worked that morning, his CB radio came to life. The voice described the smoke rising near Five Mile Tower east of Minong and reported that someone had called in a report about a campfire that had gotten out of control. Dialing that number had activated the Minong Fire Bar, a system installed by the local telephone company that caused phones to ring in thirty-three homes at the same time. All the volunteer firefighters who answered simultaneously heard the frantic voice of someone at Pappy's Tavern breathlessly reporting that a large fire had started about ten miles west of Minong north of Highway 77. Smokey rushed to his Dodge Ram Charger and headed out, picking up Minong firefighter Bobby Johnson on the way. Johnson, all of five feet two inches, took over at the wheel. This allowed Smokey to work his CB radio, study maps, and communicate with other fire departments as they arrived on scene. It also allowed him to leave the truck and shag down DNR rangers to coordinate fire suppression. The DNR had no CB radios, and the fire department trucks did not have the DNR high-band radios.

Without the ability to talk on the radio to any DNR ranger, Smokey immediately got started directing fire trucks to locations where they were most needed. He would stay on duty as "fire department boss" nonstop for thirty-six hours, working closely with DNR ranger Bill Scott, their trucks constantly maneuvering amid the flames and smoke.

Volunteer firefighters position their trucks along the road, prepared to fight the fire.

Wisconsin DNR file photo

Fire Headquarters Springs to Life

It was rapidly becoming obvious that this fire was going to be a monster, and the weather was no friend.

The fire rolled and boiled, sending towering columns up into the hazy sky over the woodlands of Wisconsin. The smoke rose three miles into the heavens and could be seen from fifty miles away.

Hudson Bay would play a huge part in the weather that day. The body of water, still frozen in April, is 1,000 miles north of Minong and larger than the eastern one-third of the United States. Splitting Canada in half, the giant bay is 700 miles east to west and 1,200 miles north to south.

A so-called Hudson Bay High had been parked over Minnesota and Wisconsin for the previous week. The combination of high temperatures and low humidity with no rain was caused by the massive ice-covered bay, producing high barometric pressures that reached across the Great Lakes. The most dangerous result was that there was no humidity recovery overnight. Instead of cool, damp evenings, the dark of night was dry as hard toast.

Then the huge system started to move, slowly rotating clockwise. First from the east it crept, then from the southeast, and on to the south and picking up speed. On Saturday, it was expected to blow hard and dry from the southwest, with no rain in sight. This system and the drought that gripped the countryside produced tinder-box conditions all afternoon and night.

In the face of these conditions, the blaze quickly grew to what firefighters call a project fire. The situation demanded that a fire headquarters be established quickly to provide a central location for fire bosses, workers, and equipment.

John DeLaMater designated a grass field on the corner of Highway 77 and Thompson Bridge Road, eleven and a half miles west of Minong, as fire headquarters; it had ample parking, good accessibility, and room for hundreds of people.

DeLaMater and Chuck Adams would work in this field, administering the largest project fire in their lives, for eighteen hours before moving the headquarters to the Minong town garage for an additional four days. There were no buildings, bathrooms, tables, or chairs at this fire headquarters, only dedicated rangers with radios, maps, and knowledge gained from training, supported by a mass of volunteers and equipment. The overhead team worked from the front seats of their trucks and studied maps on the engine hoods.

Just a day earlier, this lonely area along Highway 77 was a grassy field, dotted with the occasional jack pine. Now it was the hub of activity. It quickly filled with trucks, school buses, dozers on trailers, dump trucks, tanker trucks, fuel trucks, and hundreds of people milling around or in long lines, waiting to sign the employment register. Hundreds of backpack spray cans and shovels from ranger stations all over northwest Wisconsin stood in rows on the ground, ready for distribution to citizen firefighters. The tools were painted all the colors of the rainbow, denoting the DNR station of ownership.

Droves of people traveled to this place, hundreds arriving every hour. More than nine hundred came in the first four hours, clogging the road with traffic in both directions. Some citizens were eager to get to the fire but were frustrated with the wait in line to sign the register.

Fire boss John DeLaMater at headquarters.

Wisconsin DNR file photo

Within a few hours, fire headquarters was bustling.

Wisconsin DNR file photo

Some just grabbed a shovel and headed out in their own vehicles in their hurry to help. Those who did sign in were eligible for payment at the going rate of $2.30 per hour. If a person was assigned as crew chief, the pay leaped to $3.00 per hour.

To the untrained eye, the fire headquarters looked chaotic. In fact, the headquarters personnel worked quickly and valiantly to coordinate citizens and DNR personnel in a buildup of people and equipment unprecedented in the history of Wisconsin firefighting.

With the fire headquarters established, John DeLaMater needed to designate a crew of the most qualified and experienced personnel to manage the fire's many demands.

First he would need a boss for H Division. He assigned this, the most dangerous job on a large fire, to Minong ranger Bill Scott. The H Division boss would follow the moving head of the fire and, using dozers to pinch the sides, try to direct it toward places it could be slowed or stopped. He would use maps plus his own knowledge of the area to find locations with hardwoods instead of pines (because crown fires usually do not race through the tops of oaks and maples) and lakes and swamps where the fire could be slowed or its direction changed.

Next DeLaMater made area ranger Tom Roberts the fire's line boss. When heavy units arrived, Roberts would assign them to rangers in

H Division boss Bill Scott (in white hardhat) talks with other fire workers, including dozer operator George Becherer (seated, facing Scott).

Wisconsin DNR file photo

the divisions. When rangers checked in to fire headquarters from distant stations, Roberts would study the maps and open a new division on either the left flank or right flank of the fire. Roberts spent much time at the headquarters field, but he was also driving all over the fire ground, checking on progress, assigning new divisions, moving men and equipment.

With winds gusting and blowing stronger from every direction, fire boss John DeLaMater shivered at the awesome responsibility on his shoulders. He looked skyward at the smoke columns gaining in intensity and then got on the radio.

Line boss Tom Roberts analyzes maps in his truck.

Wisconsin DNR file photo

He and DNR forester Chuck Adams had been together all that morning, patrolling the sand pine regions west and northwest of Minong. Now they were at war. A battle loomed, against an inferno being pushed by steady, strong twenty-mile-per-hour winds blowing hard and hot. Now DeLaMater assigned Adams to the position of plans boss. Adams would gather intelligence on the speed and direction of the fire and advise the fire boss on tactics during each mile of the advancing flames.

During its first hour, the fire had already leaped forward one mile. The rate of spread of the Five Mile Tower Forest Fire—over one mile every thirty-five to forty minutes—was almost unprecedented in the annals of Wisconsin fire history.

Ralph Mortier was the Spooner DNR forest-management staff specialist, and he knew the forestlands in the area well. At fire headquarters, he was named the fire's intelligence officer. Now he was using the hood of his car as a place to spread out his forestry maps. Anchoring his papers with rocks, he plotted the path of the fire, taping maps together as the fire advanced and marking on them the burned areas. Four DNR employees served as scouts, radioing coordinates back to Mortier. Later Mortier would assess damage estimates and plan for any future possible salvage logging.

Intelligence officer Ralph Mortier (second from left) tracks the Five Mile Tower Fire with the help of scouts.

Wisconsin DNR file photo

Fire boss DeLaMater assigned a service boss, Dave Ives, who was the area fisheries manger in his day job. Ives took care of supplies, communications, law enforcement, and safety. His major undertaking was to provide equipment to fight the fire, so he picked another DNR employee at headquarters, Ed Rau, to be his equipment and personnel assistant. Ives kept track of manpower, sent people out on the fire, provided transportation, and was in charge of providing meals and water for the entire fire zone and at headquarters.

A godsend came the firefighters' way that day when a large black car drove into the headquarters field and parked on the grass next to the DNR pickup trucks. The owner of the Farmers Independent Phone Co. of Grantsburg got out and offered the fire bosses the use of his car phone. Few people had mobile phones in 1977, and DeLaMater, Ives, and others at headquarters were amazed at the gadget's efficiency. The fire leadership would use that car phone all afternoon and into the night. It proved to be an essential tool for the headquarters team and sped up the process in communicating with the outside world.

There were other happy surprises on that dreadful day. Dave Ives was grateful when the Minong Volunteer Fire Department Auxiliary showed up to provide sandwiches, Kool-Aid, water, and coffee for the firefighters.

Dozens of other individuals performed vital roles at fire headquarters. John Borkenhagen, manager of the Hayward DNR nursery, was supply officer, maintaining summaries of the location of all items sent to the fire zones and assisting in the distribution of men, trucks, dozers, and hand tools. While he was busy sending out scouts and staging crews at headquarters, he had no idea that his sons, John and Tim, both students at Hayward High School, were out on the fire lines laboring with backpack sprayers and shovels.

The fire's communications officer, Tom Teske, established and maintained communications and provided spare radios, troubleshot radio problems, and later set up temporary telephone lines.

One of the law officers, Ralph Christiansen from Brule, was in charge of all matters of law enforcement including setting up roadblocks, preventing illegal activities in the area, and providing manpower to help citizens evacuate the areas in the path of the flames.

Responsibility for the safety of everyone fell to a safety officer (whose name does not appear in any records from that frantic day). He arranged for ambulance support, dispensed first aid supplies, and gathered details on any injuries.

Many school buses were utilized during the next hours and days. Emmy Herdt was the first bus driver on scene, with a new sixty-passenger school bus owned by the Fraatz family of Minong. Jerry Fraatz's drivers were familiar with the roads in the region, and his bus garage was in nearby Minong. Herdt loaded food, beverages, shovels, back cans, axes, and other equipment through the rear emergency door; then crews of firefighters poured into the front doors, and Herdt drove them to the fire scene. As she drove back to headquarters after one shuttle trip, flames were shooting over the bus.

Other Fires Were Burning

That Saturday morning, Jim Miller's telephone was ringing off the hook in his office in Rhinelander. Miller was the Wisconsin DNR North Central District fire and recreation specialist, in charge of seventeen ranger stations and their rangers whenever there was a fire.

He knew the fire area well. He had been Hayward area ranger before John DeLaMater's arrival. Now he was ready to help in any way he could. As soon as it was determined the fire was out of control, it was designated a project fire, unleashing men, trucks, bulldozers, and equipment from all over the northern part of the state. When DeLaMater started asking for additional units in the Hayward area, other heavy units and rangers started filling in behind, protecting the areas of the stations that had sent all of their assets to the fire, including Spooner, Minong, Hayward, and Brule. Already that dry spring 130 fires had occurred in the Hayward area. And at least 56 other fires were being fought that same weekend in many northern Wisconsin locations.

Jim Miller helped organize a massive movement of personnel and equipment in support of the Hayward area. He sent rangers and some heavy units to Park Falls, Glidden, Mellon, Prentice, and others on a westward migration to protect the forestlands vacated by units fighting the growing fire west of Minong.

Miller couldn't help but think about all of the training he and the other rangers in the state had received over the previous ten years. At the heart of that training was a handheld tool called the Martini Wheel, developed by Bill Martini, the DNR fire training officer located in Tomahawk. The Martini Wheel was a sort of circular slide rule that showed the responsibilities of the various management jobs on a project fire. One side of the wheel showed the line organization, and the other showed staff organization of the overhead team. As the wheel was turned, triangle-shaped openings revealed the list of duties for each position. Made of plastic-laminated cardboard, the wheel was small enough to carry in a pocket.

Training officer Martini and DNR chief forester Gordon Landphier had been the first to set up project fire training sessions for rangers in the late 1960s and early 1970s. They taught fire-danger ratings, developing an organized system for assessing dryness, fuel moisture, and rainfall. They got the state to encourage local forest rangers to arrest the parties responsible for starting forest fires and to bill them for the costs of fighting the fire—leading directly to a reduction in the number of fires. People living and vacationing in rural northern areas recognized that the DNR was serious about careless behavior leading to fires.

Every winter rangers headed to Tomahawk for project-fire training, including daylong simulation sessions. Martini and Landphier took great pains to make these mock fires as realistic as possible. As the practice fire spread, they used maps and radios to connect the management team in one room and the organizational team in another. They simulated scouts checking on the fire progress, used real travel time, called upon aircraft to be the eyes from above, and requested men, heavy equipment, backpack cans, shovels, food, water, fuel, and other necessities. This training included all aspects of fire management, helping rangers get to the theoretical point where the

DNR railroad fire prevention specialist Jim Miller (left) shows author Bill Matthias the Martini Wheel.

Wisconsin DNR file photo by Jolene Ackerman

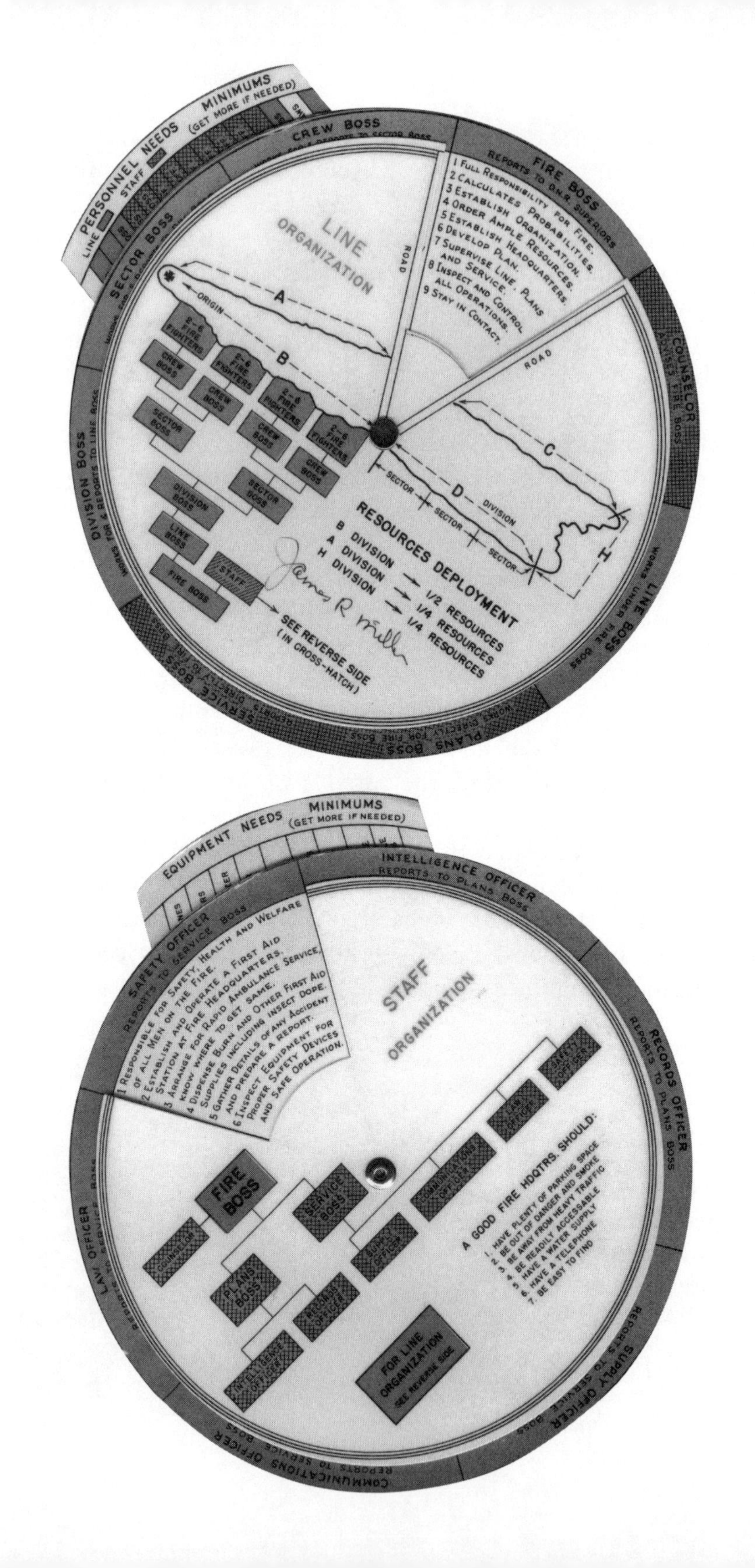

fire's head can be contained and stopped at some natural break—a different fuel type (a change in surrounding tree species and brush) or body of water.

All this may have sounded easy in the training room, working at clean, dry tables with a cup of coffee and sweet roll nearby and a bathroom around the corner, but the real thing was another matter. All that training was being tested now.

As he thought about those long days of training, Miller was on the phones keeping track of the fire's progress. He had direct lines and radio hookups to his seventeen rangers in the field. He knew that by the end of the first hour of a project fire, many of the rangers who had traveled twenty to forty miles would be arriving with equipment and supplies, clogging roads and filling the field next to Highway 77.

Miller had taught the young rangers about the Hudson Bay High and its ugly implications for fire. He knew that this was the weather 125 miles away in Minong, and he cringed as he thought about the battle ahead.

Opposite page: The DNR used this Martini wheel as a field guide to fighting large fires.

Ranger Sets a Record

At noon that Saturday, Ed Forrester was at home with his feet up. He was the ranger at the Webster DNR station, twenty miles southwest of the Five Mile Fire Tower. Ever since the University of Minnesota forestry graduate had been hired ten years earlier, he wanted the toughest assignments in the most fire-prone areas. Forrester was lean, tough, ready to meet challenges, and a good communicator. His high energy level and can-do attitude would serve him well that day in late April.

By the time the fire was put down, Forrester would have worked a total of sixty-four hours without sleep. He would be in charge of five divisions on the flanks of the Five Mile Tower blaze, more than any other ranger in Wisconsin's history.

Forrester, an avid student of fire behavior and history, knew that eighteen years earlier—on the same weekend, May 1, 1959—there was a huge fire near Grantsburg in Burnett County. The Wisconsin Conservation Department (WCD), forerunner to the DNR, had published a small green booklet, *Fire Organization Handbook,* in March the year before; nevertheless, the West Marshland Fire burned 17,560 acres and was fought without much organization, planning, or training by the persons battling the blaze. The general disorganization caused the WCD to come up with better ways of fighting large fires.

The WCD was rusty when it came to firefighting. Sometimes decades passed between big fires. Drought conditions came in cycles, with ample rainfall most other years. In northern Wisconsin, a major drought occurred in the late 1950s, and another in 1976 and 1977. These long time spans lulled the rangers and fire administrators into thinking maybe there wouldn't be another big one. However, the department did conduct many training sessions and simulations of firefighting tactics.

Ed Forrester had been dispatched to a fire near Grantsburg at 7:30 Friday morning, April 29, and had worked there for thirty hours. Just before noon on Saturday, he drove home, took a shower, and had lunch, all the while thinking the morning was too quiet. That would change soon. He didn't even get a nap. The call came in at 1:30 p.m. A smoke plume had been detected near the Five Mile Fire Tower, and there were reports on private CB transmissions of a campfire jumping out of its rock ring in windy conditions. He heard that the Minong Volunteer Fire Department's telephone fire-bar system had been activated.

Fire boss John DeLaMater requested Forrester and his two forest-fire-control assistants driving heavy units with plows. They drove to the fire's location west of Minong on State Highway 77, just past the Namekagon River Bridge. Dog tired, but with a shower and clean clothes, Forrester was energized as he raced to the scene, listening to the transmissions on his high-band radio. The radio was buzzing with fire activity. Forrester had the feeling this fire was getting out of control.

With his recent lunch sitting like a lump in his stomach, Forrester was given the B Division boss position, on the most dangerous flank of the fire. It would be another twelve hours before ranger Tom Roberts could catch up with him and, sometime after midnight, give Ed Forrester some water and a sandwich.

That day and into the night, Forrester asked to try a new technique, and Roberts agreed. Forrester wanted to stay near the head of the massive flames on the right flank, scouting ahead and supervising dozers, fire trucks, and squad leaders. At various road intersections or lakes, a new division would be established and given a new letter code. The fire was racing fast, better than one mile an hour. According to training protocol, divisions should not get too big. Each division needed a ranger in charge, dozers with plows, tanker trucks with hoses, and many firefighters broken down into crews of two to six, each with a crew boss. Some divisions had four or more crews working.

Forrester was given the responsibility of setting up each new division, and in this method he leapfrogged his way north as the boss of five divisions. That would keep him near the fast-moving inferno, always trying to get his dozers to plow as close to the eastern flank as safely possible in an effort to squeeze the fire inward.

Saturday afternoon and evening, as more DNR rangers reported to fire headquarters after driving long distances from their stations in northwest Wisconsin, they were assigned to divisions on the right and left flanks and ordered to hold the fire lines and plow new lines back and forth so the fire could not break through. Ed Forrester was there for all the action.

After Ed went north following the flame front, Spooner DNR ranger Don Monson took over the B Division through the night.

Wardens Join the Fray

Minong area DNR conservation warden Ed Nelson knew every nook and cranny of his territory. He had driven down every road, driveway, trail, path, and boat landing in his years as a conservation warden, checking the guns, fishing equipment, and licenses of hunters and anglers and carrying out dozens of other tasks. He had a three-wheeler, a boat and motor, and some strong legs, and he could go anywhere in his territory, day or night. Nelson also gave of his personal time teaching hunting and boating safety classes to kids and adults.

On the morning of the fire, Nelson was out in his DNR truck, patrolling roads and boat landings, looking for anglers trying to open the fishing season a week early or anyone burning brush or trash illegally. Then his truck radio crackled to life with reports of the fire.

The fire changed Nelson's duties and those of many other wardens that day. Along with sheriff's deputies, state patrol officers, and pilots with PA systems on DNR airplanes, local wardens quickly began evacuating residents.

Nelson drove to the fire scene, assessed the fire's path, and then drove on, trying to evacuate anyone living in the expected hot region. He went door-to-door, advising people to get out and where to go for safety. He worked all afternoon and into the night, not resting until all citizens were out of harm's way. A few hours into the fire, Nelson was way out on the western edge in the Bear Track and No Man's Lake areas, because the crown fire was heading north, directly toward the homes in those locations.

Drummond warden George Phillips was called to block Highway 77, allowing only firefighters and fire vehicles into the fire headquarters zone.

Warden Ralph Christiansen, the warden supervisor in the Brule area, shuffled other wardens to places that needed roadblocks. He also sent some wardens ahead of the flames to evacuate people who had not yet left their homes.

Max Harter, the conservation warden from Grantsburg, grew up in Gordon and knew the area well. He was sent to drive up and down roads, evacuating people living in the expected path of the fire. In addition, he went back into the burned areas to watch for looting of abandoned cabins and theft of property. At one point he saw huge piles of harvested pulp logs burning in the Bill Frahm memorial forestlands along Highway T. An old porcupine was nearby, struggling to get away from the flames. Max used his shovel and gave the prickly critter a ride to safety.

Warden Joe Davidowski from Superior drove in dangerous conditions in front of the flaming head, evacuating the few people still there from homes and resorts. He teamed up with Bob Cleary, another warden, and spent all night and into the day on Sunday blocking roads.

Warden Milt Dieckman spent three days blocking traffic to Smith Bridge Road, which runs north from Highway 77, to keep unauthorized people out of the fire zone to make the road available for fire vehicles and DNR patrols.

Here Come the Dozers

One side is flaming red, inky black smoke, and hot, the other green and cooler. That's what dozer drivers experience in a fire. They drive alongside the fire, plowing deep, flat-bottomed ditches in the soil, pushing over trees, and cutting roots, bucking up and down along the way. These brave forestry technicians, heavy-unit operators, or just plain dozer drivers are trained and ready to drive into danger in an effort to encircle the flames.

No big Wisconsin fire in dry, windy conditions can be contained by men and women on foot with hand tools. Time and again, the steel-tracked vehicle pulling a heavy steel plow has proved itself to be absolutely necessary to stop fires. In the case of the Five Mile Tower Fire, many dozers with plows, and some with front blades alone, would be needed, working on the left and right sides of the racing inferno, always trying to pinch the fire ever narrower—finally to a point—and ultimately put it out.

At the start of the fire, forest ranger Bill Scott had immediately radioed fire boss John DeLaMater with some hard facts. The fire was in a difficult area, full of dry fields and pines with sandy soils, hills, valleys, and two rivers. It would be a tough place to fight a fire, with the rivers blocking clear access by road. Vehicles would have to travel long distances to find bridges, cross the rivers, and head back to the fire's flanks. Valuable time would be lost. And dozers would be indispensable.

Before even getting to the fire, Scott had called for backup units, including the Minong dozer parked at Webb Lake. George Becherer jumped in his two-ton truck pulling the trailer with the dozer and plow chained down. Scott called for four more heavy units and two additional rangers.

In 1977, most of the DNR dozers were the T-6 type, small and underpowered, made by International. These dozers pulled a sulky plow, similar to a farm field plow, with two wheels similar to the old sulky buggies pulled by horses in years past. Most of the plows had steel wheels and were connected to the dozers by a hitch system. The state owned only a few new John Deere 450 dozers—and these proved to be superior in every way. The plow behind the JD 450 was connected directly to the rear of the dozer and moved up and down hydraulically. The JD 450 was much more powerful than the previous generation of dozers. It was a great improvement, and ranger stations that had one considered themselves lucky. At the time of the Five Mile Tower Fire, there were only two JD 450s available, one at the Minong DNR station and one at Webster.

Almost immediately after the fire started, dozer operators were on the scene. As Pete Paske arrived with one of the JD 450s and plow, landowner John Schultz ran up, grabbed a five-gallon backpack can and shovel, and yelled to Paske, pointing to where the fire had started. Paske began cutting furrows in the sod and woods of the 132-acre Schultz property in an attempt to stop the advance and put a dirt ring around the flames.

As Paske was plowing near the Totagatic River, his hydraulic pump blew a seal and started spraying fluid. He had to leave the fire and get the unit fixed. An hour later, he returned and plowed with Spooner units doubling and tripling the lines, constructing drivable firebreaks. Each firebreak was a six-foot-wide dirt path in the woods, one or two feet deep and two feet wide at its flat bottom to allow firefighters on foot to walk along carrying their heavy equipment. After two or three fire lines are plowed, a heavy dozer would come behind, smoothing everything over so four-by-four pickup trucks could drive on the firebreak and patrol the fire's edges. Paske would plow all night.

Meanwhile, George Becherer arrived with his Minong heavy unit, a newer six-cylinder International Harvester T-6. He started plowing deep furrows into the sandy soil on the west side of the fire. Flames were rushing north with the wind like a kite, dancing back and forth, sending sparks in all directions.

When DeLaMater heard on the radio of the fire's rapid spread in the first half hour, he called for every ranger station in his area to travel to the fire, and within minutes, trucks, trailers, and dozers with plows were racing toward fire headquarters from Webster, Grantsburg, Hayward, Winter, Ladysmith, North Spooner, South Spooner, and Gordon.

In the second hour of the fire, Everett Hanson, heavy-equipment operator from the Grantsburg DNR ranger station, arrived at fire headquarters in a two-ton truck hauling a new John Deere 450C. He unloaded and was instructed to start on the B flank, the beginning of the fire on the right side. Sometimes a fire can be contained within the B zone—but not that day. B Division boss Ed Forrester was attempting to keep the fire on the south side of the Totagatic River. When Hanson reached the bluff overlooking the Totagatic, he saw Becherer, who said he would plow down to the river's edge if Hanson promised to wait and pull him back up again with heavy chains. It was a dangerous situation, but the fire line needed to be tied into the river. In the process, Becherer's dozer was disabled, leaning on one track against a tree. Hanson got him going again by pushing and shoving with his machine.

Soon Hanson came upon a burning dozer. Carl Backhaus and Bruce Williams, dozer-plow operators from the Spooner Ranger Station, were there with the Spooner heavy unit, a four-cylinder International T-6, with a fire burning in its belly pan caused by dry pine needles catching fire. Hanson quickly put it out with his hose and then continued plowing into the flames on the eastern edge of the fire. He would spend most of the night and all day Sunday adding ditches all over the fire ground, working to near exhaustion. Although food and water were pretty much unavailable, whenever he called on the radio for fuel it would be delivered promptly.

He finally made it back to Grantsburg about midnight on Sunday. The night before the Five Mile Tower Fire, he had plowed on another blaze. By the end, Hanson had worked sixty-three hours without sleep.

Don Crotteau was one of the operators out of the Webster Ranger Station. Early that Saturday Crotteau drove an IHC T-6 Cletrac with a sulky plow on the fire's B Division. Later he worked in H Division with Bill Scott, alongside the crown fire. He helped George Becherer plow a line on the east flank between Highway T and Cranberry Lake that was later credited with keeping the fire pinched on the western side of Cranberry Lake. Crotteau and Becherer also helped save View Point Lodge, just east of County I on the Minong Flowage about a mile south of County T, and Kresch's Harbor bar and restaurant located on Highway T at the Cranberry narrows bridge, along with property along many lakes to the east. Crotteau put in six miles of firebreaks on the fire.

At one point he got caught in some heavy slash and downed trees and brush and needed another dozer to help him get out. At one heavily burning section of the fire, the hydraulic hose broke on Crotteau's dozer, spraying superheated fluid over the engine, which was covered with dry pine needles. The T-6 started on fire, but Crotteau was able to put it out using his own hose, nozzle, and water from his seventy-five-gallon tank.

Later, plowing near Deer Lake, Crotteau saw the fire nearing a cabin and boathouse. He radioed Bill Scott at headquarters, and soon one of the Minong fire trucks drove into the yard and saved the structures.

After daylight on Sunday, having had no food all night long, Crotteau was famished. He was on a sand road when he saw a woman driving up in a car in the middle of the fire. She asked him where all of the firefighters were because she had a covered bowl of hot dish along with several spoons. Don crawled down from his dozer and enjoyed his first meal in nineteen hours, eating right out of the bowl. He thanked the kind lady, climbed back on the T-6, and resumed plowing.

Late Sunday afternoon the wind shifted, threatening the eastern flank. Because Crotteau and Becherer were close by, they and other heavy units were able to run down the flame-up and put a line around it. Again, Crotteau blew a hydraulic pump and couldn't lift the plow. But he needed to get across a road. He drove the dozer across, blade down, to continue his pressing assignment.

It wasn't until Sunday night, about dark, that Crotteau would get back to headquarters and be released to go home to Webster. He had been on duty for sixty hours without rest.

Allen Zaloudek arrived at fire headquarters in his two-ton truck pulling a new John Deere 450 with no plow, just a front blade. This dozer was used for building roads and in forestry operations. He arrived forty-five minutes after the fire had started and was assigned to the B Division with ranger Ed Forrester. The fire was moving rapidly on the right flank, still on the south side of the Totagatic River. Zaloudek worked with Pete Paske and George Becherer at the start of the fire but got stymied when the fire reached the river. His dozer could not cross the wetlands and river, so he had to waste valuable time driving the half hour back to his truck and trailer to load up and drive around the fire on a bridge. About an hour later, he was back making the fire lines wider.

As the fire leaped the river, it blasted forward and became even more dangerous. The flame front had hit the jack pine area, with miles and miles of dry trees in every direction.

Soon line boss Tom Roberts set up the D Division and Gordon DNR ranger Barry Stanek was assigned as boss. Zaloudek followed the fire on the right flank and continued pushing trees and brush to make wider fire lanes after the first dozers with plows had gone through. Fire teams were worried about an expected wind shift. All afternoon and through the night, Zaloudek's and one other dozer dug a twelve-foot-wide drivable road in the flat, sandy country. The fire had long since passed by and was roaring forward, but the one-and-a-half-mile D Division had to hold and not be breached by a wind shift—or the right edge of the fire could become the head and roar all the way to Minong.

The flames towered twelve feet over Zaloudek's dozer. Erratic winds fed the fire's insatiable desire for oxygen. Once Zaloudek almost was trapped and had to back up into the black zone of already-burned grass, brush, and trees for safety. He could feel the heat and hear the fire, loud as a locomotive over the roar of his JD 450 engine. All night long, he and Stanek worked to hold D Division. They would work nonstop until 6:00 p.m. Sunday, a total of thirty hours.

DNR Northwest District assistant director Bob Becker spent the first night of the fire in the outer office at the Spooner dispatch center, manning the busy phones while Barb Raasch handled dispatch. He was calling for help to assist the rangers with equipment needed at fire headquarters. He also served as the media officer for the fire, fielding questions from the press and from citizens. His primary job began when DeLaMater asked for more backup from heavy dozers—those owned by excavation and road building contractors and by county governments for clearing land and building roadways. Becker spent hours on the telephones, trying to bring in every available piece of equipment. Many of the contactors worked late and were hard to reach in those days before cell phones.

Becker also contacted the Madison DNR state offices, which had set up a command center for the fire, and asked Governor Patrick Lucey to authorize the National Guard to release its big D-9 Caterpillar dozers to the Five Mile Tower Fire. The request was approved, and that afternoon many eyewitnesses watched the massive, olive-drab dozers with their huge blades roar and snort through the woods, trees crashing on all sides.

In the midst of his equipment search, Becker was fielding other calls. Newspaper editors and radio and TV news directors wanted statements about the fire, its cause, and how many acres were being destroyed. Media and citizens called asking for reports on deaths or injuries and the number of homes and cabins destroyed. Becker fielded the questions as he tried to find enough large dozers to send up to the fire scene and headquarters twenty miles away. He ordered twenty-five private contractor, National Guard, and county dozers to supplement the DNR's ten heavy units.

All the while, Becker was worried about reports of an expected wind shift that would leave the entire ten-mile-long eastern flank exposed to a strong sideways wind. It was critical to beef up this area with heavy dozers. This would be night work. Becker's first choices were dozers with lights, but in 1977 not all dozers had them. It could be terrifying for a firefighter on foot to hear a snarling dozer knocking down trees and not be able to see the machine. The operator cannot see outside of his cab, and people fighting fires on foot cannot see the big dozer.

Jim Stordahl, DNR Douglas County forester, learned how *not* to flag down a dozer that night. He was leading twin giant D-8 bulldozers owned by the Olson Brothers Excavation Company from north of Solon Springs. The Olsons were experts on road building during daylight hours and were skilled at knocking down trees. But neither of these dozers had lights. It was extremely dangerous leading caterpillars with no lights at night. They roared through the darkness, pushing over forty-foot jack pines like they were matchsticks. The decibels rose with the gears grinding, the steel tracks screeching, the tracks clattering against the stumps, and downed trees crashing. Stordahl had one flashlight, trying to stay ahead of the equipment, running and dodging falling trees as he guided the operators in making wide fire lanes. The sky was aglow as if it was the end of the world, but the machines were invisible in the black forest.

When Stordahl wanted the operators to stop, he waved his light frantically, but they kept on coming, crashing and growling through the trees and underbrush. One jack pine fell dangerously close to Stordahl. He turned out his light and ran sideways and then back to yell at the drivers. They stopped. When he breathlessly approached the idling monsters, one of the operators called down, "When you want us to stop, just turn out the flashlight—don't wave it!"

Dozers loaded on trailers line the road, waiting to be deployed to the Five Mile Tower Fire.

Wisconsin DNR file photo

After the Five Mile Tower Fire, criticism rained down on the DNR officials, with stories claiming that bulldozers were idling on Highway T near the eastern flank of the fire, others on Highway T on the western edge, and still more to the south, parked along State Highway 77. Because of the lack of cooperative training with the heavy-equipment owners and operators and limited knowledge of crown fires, even the equipment operators themselves fueled these criticisms, complaining about having to wait and not getting to the task at hand. People with little knowledge of fire management speculated that the crown fire could have been stopped if the dozers had been unloaded and put to work earlier.

What the critics did not know was that a hot crown fire cannot be stopped at the head by any means. Additionally, dozers were being held in reserve for emergencies later and to push trees over and build miles and miles of wider fire lines all night long. If all of the equipment was put on the fire, where would the fire boss be with no dozers in reserve for emergencies and flare-ups? What would happen if the wind did shift, threatening the fire's right flank?

Sandwich Queens

Jerry and Kathy Fraatz of Minong owned all of the school buses that transported the four hundred kindergarten through twelfth-grade students of Northwood Schools. They also owned Fraatz Trucking and had heavy equipment and a gravel pit for landscaping and construction projects.

Jerry Fraatz was on the Minong Fire Department. On this Saturday, Jerry and his father, Herman, were working on buses in the garage, and Kathy was cleaning the house and doing the business bookwork, catching up from a busy week. They were looking forward to a relaxing weekend, but it was not to be. In a few hours, Jerry would be using his dozer to dig fire lanes on a huge forest fire, Kathy and some friends would be making sandwiches for fire workers, and the Fraatzes' school buses would be hauling workers to the flames.

There were hungry firefighters to feed, so Kathy Fraatz called Nancy Block and other friends to help make hundreds of sandwiches. Kathy Fraatz was a member of the Minong Fire Department Auxiliary, and soon other members showed up. From the Fraatz kitchen, they watched the billowing clouds of smoke ten miles west and shivered at the thought of the people trying to put out the fire. The Fraatzes had a base station CB radio, so Kathy and the other workers in the kitchen could hear the communications of the volunteer fire departments and many times the voice of Chief Smokey Smith.

Later that evening, the Minong Auxiliary members, including Jean Legg, Audrey Wallace, Barb Wetzel, Bev Love, Karla Smith, Sharon Smith, Teresa Sikorski, Janice Link, Marg Waggoner, Linda Henson, and Donna Sybers, would move to the more spacious fire hall and adjoining Minong Village Hall to continue preparing food and coffee. They called themselves the "Sandwich Queens of Minong." All they had to do was step outside the Minong Fire Hall and look west to see the orange sky and red horizon. Most of the villagers were fighting the fire, and the Sandwich Queens had to feed the multitudes.

The women called Dewing's Super Value and Link's grocery, both in Minong, and ordered everything they could for sandwiches. Freshly sliced bologna, ham, sausages, and cheese were delivered, along with truckloads of bread loaves. The Link family stepped to the plate that day, keeping the meat department open and the store available all afternoon and into the night.

Dick Scott, who worked for the Link family, stuffed the back of his large Ford station wagon full of the sandwiches and drinks and then drove out Highway T to get near the firefighters. He gave the police and wardens food along the way. He went down sand roads in every direction looking for hungry men and women working on the fire. Later he would witness the fires burning on the west side of Person Lake and between Crystal and Cranberry Lakes. He worked all afternoon and all night.

The sandwiches, Kool-Aid, and coffee were many local women's link to the fire and those they loved. Among them was Nona Yrjanainen. She and her husband, Waino, were Finns living in Minong. They owned Nona's Café on Highway 53. When Nona saw the smoke in the distance she called her kitchen helpers into the restaurant. After working all day, she went to the Minong Village Hall and helped make sandwiches until 3:00 a.m.

Due to a coffee shortage in 1977, the price was at a premium and supplies were limited. But Nona Yrjanainen knew where to find some. A family that lived in the path of the fire had temporarily moved in with Nona and Waino in Minong. The couple had been hoarding coffee beans in the shortage. Their entire automobile trunk was filled with coffee.

After working through most of the night, Nona opened her restaurant at 4:00 a.m., and immediately the place was busy. Most of the firefighters were headed home for some sleep. Some were going immediately back to the fire to continue working. The coffee was flowing freely.

H Division boss Bill Scott's wife, Kathy, knew from experience that large fires take a long time to control. She, too, gathered a group of women at her home and made hundreds of sandwiches.

The auxiliaries of the Solon Springs Volunteer Fire Department and Gordon Fire Department also made sandwiches by the hundreds. In Solon Springs, Gale Larson, Mary Hill, Barb DeFore, Molly Lucas, and many others shuttled to and from the local grocery store. Jerry Larson, who worked at the store, was given a key by the owners to obtain supplies. The Gordon Auxiliary worked out of the Gordon Town Hall, where people arrived throughout the afternoon and evening to prepare the food and drink. Grocery stores were emptied in Spooner, Minong, and Solon Springs in order to feed fourteen hundred citizen firefighters and hundreds of DNR personnel and volunteer firemen.

While Dick Scott delivered his supplies, women took pickup loads of coffee and sandwiches to a staging point after dark at the Five Corners intersection, three miles north of Highway T on Crystal Lake Road. Men, women, boys, and girls came from all directions for sustenance. Trucks, cars, tankers, and pumpers were parked all around. The Sandwich Queens from Gordon, Minong, Wascott, and Solon Springs had come through.

It was cold out there in the dark, and the wind was still blowing at almost twenty miles per hour. Firefighters, many of whom had not eaten in seven hours, sat exhausted near their vehicles to get some heat from the engines and dry their sweaty, soot-caked T-shirts.

Some of the women serving food were shivering. Warden Bill Hoyt opened his trunk and dragged out every piece of clothing he had available. He gave the workers his wardens' winter jackets, vests, coats, and even rain gear. They continued serving food for hours into the night as the horizon glowed from flames flickering above the treetops.

After the fire roared past View Point Lodge, a resort and tavern right on the edge of the burned area on Highway I just south of T, owners Dottie and Bob Fries set up sandwich lines on the bar and called on their friend Carol Barbee to help. They went to Link's store and got bread and meat and whatever they could grab to make lunches for the firefighters. All weekend they gave out fuel from their underground gas tanks as well as candy bars, chips, and pop to hungry firefighters. When the fire was over, almost every person came back to pay for the gas and snacks. The Frieses appreciated the northwoods honesty.

Teenage Firefighters

He was only seventeen, but high school junior Bud Schaefer knew he loved this work. He had volunteered to join the Northwood High School primary fire crew and had become one of the leaders of the elite group. Most days during this dry spring, Schaefer and the rest of his crew had been coming to school dressed in jeans and their most comfortable outdoor leather boots. Their fire-retardant shirts hung in their lockers in case the team should be called to action. But on Saturday morning, April 30, Schaefer arrived at the ranger station tired from only four hours of sleep. He had been up late the night before, netting smelt with friends forty miles north, on the shores of Lake Superior. This annual ritual had yielded a large cooler full of fresh smelt. Schaefer's plan was to clean them Saturday afternoon after getting home from his job on standby at the Minong Ranger Station. Instead, he would be fighting the forest fire of the half century and would not return home until thirty-six hours later.

Northwood High School had a primary crew of ten students and a secondary crew of the same number. One requirement for primary crew members was that they not be involved in a school spring sport, so they would be available any day or night. Schaefer was one of the primary crew members who would eventually make firefighting with the DNR his life's work. It all started when he was selected to work weekends on standby at the Minong Ranger Station. When Bill Scott asked him to get in his truck to race to a fire west of Minong that day in April, Schaefer was ready. The smelt were forgotten.

That Saturday, Northwood junior Mark Radzak, seventeen, was working at a part-time job putting together boat trailers for Doug Denninger, an art teacher at Northwood Schools. After their lunch break, both saw the plume of smoke eight miles to the west. Muscular, with thick brown hair and sideburns, Radzak looked older than his age. He loved the woods and was always hiking or driving around, trying to find new places to fish or hunt. He knew the territory better than most men. He couldn't wait to get on that fire and asked to take off the rest of the afternoon. He jumped on his motorcycle and raced westward on State Highway 77.

Soon he arrived at Thompson Bridge Road, where he was startled to see several DNR pickup trucks and bulldozers in a grass opening. Radzak was also shocked to see fifty or sixty people milling around, waiting to sign on with someone and go fight the fire. He parked his cycle, bypassed the signup lines, and grabbed a shovel off one of the tractors. He met up with fellow Northwood High School fire crew members Scott Clark and Jim Barrett, both sixteen, and the boys headed out.

They walked toward the smoke on the east flank of the rising firestorm. They knew from training on the student fire crews that the right flank of a fire was always the most dangerous because of the earth's rotation and something the rangers called the Coriolis effect. In his many hours spent at the ranger station, Mark had often overheard the rangers talking about the dreaded crown fire and how the earth's spin always encourages the fire to bend to the right. He wanted to be on the tough side of this northbound fire train. He got his wish when the trio caught up to a DNR dozer, bucking and snarling as it plowed a two-foot-deep ditch amid the stumps, trees, roots, and bush clumps on the fire's eastern flank.

Radzak fought behind this dozer all the way to Little Sand Road. Later he ran into fellow student Keith Postl and equipment operator Bob Hoyt, from Gordon. They all came upon a red pine plantation burning fiercely, belching smoke so thick they couldn't breathe. Radzak and Postl, who were on foot, had to lie flat, stick their noses into the cool soil, and breathe air near to the ground as the flames went over them in the crowns of the trees.

Later, Radzak and Scott Clark came upon an abandoned army surplus six-by-six water truck with pump and hoses. They jumped in and drove it down freshly cut furrows, breaking down small trees on the way. A ranger directed them to flare-ups needing water. The boys took turns, alternating driving and sitting on top, spraying with the nozzle. On Sunday, after fighting next to the flames all night, they were in the northwest sector of the fire on West Mail Road in Wascott when their truck broke a gas line. A DNR mechanic showed up to fix the truck, questioned their ages, and took the vehicle away from them. The boys finished the fire on foot. Radzak worked for thirty-three hours before returning to his motorcycle and heading home for sleep and his mother's cooking.

Northwood High student Blaine Peterson, seventeen, had been hired at the Gordon Ranger Station for weekend firefighting assistance along with fellow student Keith Postl. They both wore jeans and leather boots and were issued yellow fire-retardant shirts. When the fire call came in, Peterson was assigned to ride with ranger Barry Stanek. They were quickly on the road, heading for the fire headquarters. At the fire scene, Peterson dug fire lines by hand behind a John Deere 450. He stayed with Stanek most of the afternoon and then spent many hours working the eastern edge of the fire in the black areas already burned. When night descended, he was lost for a time, following dozers and working with several crews, not knowing where he was. Then he got back with Stanek, who told him to stay with the truck while Stanek hiked into the burn to make an assessment.

A group of citizen firefighters came out of the woods. They were dirty, tired, cold, and hungry. They wanted to get back to fire headquarters. Peterson ferried them back and then returned to wait for his boss. Finally Stanek walked out, and they drove off in search of food and water. Both had worked from 8:00 a.m. until midnight without food. They finally caught up with someone bearing sandwiches. The sandwiches had but one slice of bologna in them, but they tasted delicious to the famished firefighters.

Earlier that day, at about 4:00 p.m., Peterson had found himself working in the dusty, black ash, putting out small fires still burning in stumps and brush piles. He had been without water for seven hours and needed to fill his backpack sprayer. A dozer stopped and filled the can to overflowing. Peterson took his hard hat off and filled it with the rusty, foamy water coming out of the dozer's tank. He drank it down. That dirty water quenched his thirst, but he was lucky he didn't get sick, because the water came from a muddy stream and an old steel reservoir.

Dirty water wasn't the only liquid the high school students drank that night and the following nights. Kay Wanless, who owned Wanless Tavern on Crystal Lake, was so appreciative when her bar was saved from the flames that she provided cases of cold beer and pop to the crew.

D. J. Aderman, fifteen, was on the trained student fire crew at Hayward High School. He was working that day as weekend help at the Hayward Ranger Station. D. J. was young, tough, and eager to get to the flames, but he was not called to go with the initial group because of his age, which upset him. Later he rode to fire headquarters with ranger Tom Quilty, excited to finally be getting to the action. But his heart dropped at headquarters when DeLaMater and John Borkenhagen assigned him the important task of sitting at a card table and signing in the citizen volunteers. D. J. was angry. He wanted to be out fighting the fire but was assigned a desk job, important though it may be. Nevertheless, he did his job as best he could for hours. As he wrote down the names of the folks in line, he could hear the radios crackling with messages about the fire. Some of the reports were calm and measured; others were excited, the voices yelling about crown fires, heat, danger. He wanted to be among them.

Finally, later that evening Aderman was sent to the western, or left, flank of the fire, where other Hayward students were working behind the flame front, keeping the lines from jumping. All night long he labored with his friends. They worked hard, with fifty-pound backpack cans and shovels, and he was never so tired. The boys were wet, sweaty, and dirty as they sprayed and shoveled sand on the burning stumps and tree trunks. All night long Aderman could hear the roar of the crown fire in the distance and see the glow in the sky. He knew the rangers had put the student crews in areas of little danger—and he could understand their decision after hearing the freight train sound of the crown fire miles away.

Aderman and his friends kept working, for thirty-seven hours without stopping, until they were exhausted and allowed a nap on dry pine needles sometime Sunday afternoon. They grew up during that night of no sleep. Years later, as forester in charge of Johnson Timber in Hayward and its Future Wood Corporation, Aderman recalled that the Five Mile Tower Fire had tested his toughness and given him a renewed respect for the rangers and their passion for forest resources.

In all, more than one hundred high school students helped fight the Five Mile Tower Fire as members of the Northwood, Hayward, Spooner, and Solon Springs High School fire crews. Today, more than thirty years later, many of them are still in firefighting as volunteer firefighters, foresters, and equipment operators. Their commitment is a testimony to their grit, citizenship, love of the outdoors, and zeal to help protect natural resources.

Animals Didn't Have a Chance

The swirling ball of fire spread upward, engulfing everything in its path, heading north, out of control. The orange beast first flew through the tops of the tinder-dry jack pines, and the main firestorm followed, hugging the ground at one thousand degrees, popping, snapping, moaning, and rushing forward, hungry to destroy. Its appetite was enormous, eating grass, pine needles, pine cones, brush, weeds, blueberries, raspberries, and all varieties of underbrush. It incinerated power poles, wires, trees, cars, garages, outhouses, barns, and cabins.

Animals didn't have a chance in the voracious blaze. Unless they were lucky enough to run at right angles to impending doom and get out into the cool sides, they were eaten alive, burned while racing, digging, or climbing in frantic attempts to flee. They could hear it coming: trees crashing down, branches flying after being burned off, the swishing, howling, and rushing of leaves, needles, dust, dirt, and dry twigs all being scooped up by the horrendous gusts feeding the flames.

Flaming chunks of pine needles, birch bark, and wood rained down on all sides, each burning ember immediately growing into its own small forest fire. Within minutes, these spots of flames would suck in wind and merge into an inferno in front of the larger fire. Deer, bears, raccoons, badgers, coyotes, squirrels, blue jays, chickadees, and many other creatures of the pine and scrub oak forest were trapped. Some dropped over from suffocation even before their hair or feathers started burning. Others, caught too close, had their eyes scorched and vision destroyed. In a blind, helpless dash of fear, they ran, following their instincts to travel upwind, into the maelstrom. Sometimes they hit trees in their random darts and dodges, injuring themselves severely before being engulfed in the furnace, their skin turned black in an instant, hair gone seconds before they dropped to the ashes of the forest.

A mile ahead of the fire, a doe and her newborn fawns were sleeping in the tall grass of a former farm field. The Totagatic River was just below them, providing water to drink and a habitat of fresh young buds and twigs of the bottomlands. Their heads jerked up as they smelled the smoke, but they stayed frozen, curled up on the ground. They heard the fire approaching. Then they heard the cracking noise of branches and trees breaking. The terrified deer did not move, trying to stay hidden as the fire approached.

Birdsongs turned to shrill screams. When a lick of yellow flames flickered in the distant jack pines beyond the field, the deer jumped up in unison. Racing before the inferno, they dashed past a cabin, hardly noticing a man on his roof squirting water from his garden hose and yelling at his wife to pack the car and get it running.

An owl flew from its perch high in a fifty-year-old red pine and careened upward, out of control in the superheated updraft of soot, heat, smoke, and flaming debris. It choked, wings crumpled as though it had been shot, and plunged to the earth, striking branches on the way down. It hit the earth with a thud, lungs seared and neck broken.

Animals living in tunnels or dens dug into the cool sand where their roofs were held up by the grip of hundreds of roots in the soil above them. When they scurried from the horrific sounds and heat into their lairs, they were safe for a while. The crown flames raced high overhead. But their comfort was short-lived. Soon the mass of ground flames consuming everything in its path roared forward, and the heat cooked them alive.

The Fire Tree

The jack pine's needles are kind of ugly, twisted, short, and two to a bunch. The cones are small, pointed, closed up tightly, hard as a golf ball. This is not a gracious tree with beautiful lines, not used for ornamental plantings around homes. Unlike the stately red or white pines that reach ages of two hundred to three hundred years, this tree must be harvested by the age of fifty or sixty before it starts dying.

No, *Pinus banksiana* is not a pretty tree. Named for Sir Joseph Banks, the naturalist and botanist on Captain Cook's first voyage, this tree is known in northern Wisconsin as the jack pine. Elsewhere in America, it is called eastern jack, gray pine, black pine, black jack pine, prince's pine, Bank's pine, Hudson Bay pine, scrub pine, or northern scrub pine.

Jack pines thrive in areas with sandy soils. They need full sunlight and grow best with their own kind in large areas of forestlands. Before

Jack pine, the resilient "fire tree."

Wisconsin DNR file photo by James R. Miller

European settlement of this region, jack pines were seen in dense groves, all the same age, thriving. They were the same age because of fire.

The jack pine loves fire. When its habitat is scorched by fire, it grows back thicker than before, thanks to its small, tightly closed cones. The cones are tiny, small enough to be carried in the mouth of a red squirrel, but they are tough, filled with natural resin on their sharp scales. The resin holds the cones closed long after they mature. Unlike the cones of their cousin trees, jack pine cones remain closed until heat melts the resin. The cone will not grow a new tree unless there is a lot of heat—like in a forest fire.

Fire races through the tops of the large jack pines. This searing heat causes a chemical reaction in the resin in the cones. Often a burned jack pine's cones are still attached to the branches, looking like pointed chunks of charcoal. But slowly, over a few minutes or hours, the cones open to rain seeds onto the ash below. Some seeds die from leftover heat or glowing debris, but most of them land on cooled ashes or bare ground. All it takes is one good rainfall to dissolve the ashes, helping the tiny seeds take hold in the soil and begin to grow.

The cones at the tops of the trees release the most seeds, because the fire goes through the tree crowns within seconds. The heat is intense but does not last long, saving the protected seeds. Cones closer to or on the ground get burned and do not produce many seeds.

The Five Mile Tower Fire would blast through thousands of acres of jack pine. Mosinee Paper Company (now Wausau Paper Company) alone lost 4,426 acres of mostly jack pine. Mosinee foresters Steve Coffin and Pat Sheller later reported an amazing statistic: 65 percent of the company's acreage that was destroyed in the fire regenerated naturally. Some of the burned jack pine stands were back only a few years after the fire, with more than five thousand trees in each acre, compared to one thousand per acre if they had been planted by hand or machine. Pine cone seed production is a wonderful thing.

Birches Blow Up

By about 2:30 p.m., one hour in, the fire had burned two hundred acres and was gaining momentum, heading straight for the Totagatic River. Surely it would stop there. Flames leaped up the trunks into a dense stand of jack pines, sending orange and yellow coals flying in the wind.

Birch trees stood in their white beauty along the river, paperlike bark curling as the trunks expanded in their growth cycle. Their leaves were just coming out, tiny, light green like new grass. The birch roots were digging deeper, searching for moisture. The Totagatic was just a trickle, and the groundwater levels were dropping from the two-year drought.

As the flames from a jack pine grove roared closer, DNR dozer operator Pete Paske saw the bark on the birches catch fire. Pieces of flaming bark exploded off the trunks, burning furiously with inky black smoke, starting a thousand other fires in the dry leaves. The wind blew this mass of fireworks forward across the river as though it wasn't even there. The Totagatic was breached, and the living, breathing monster had a life of its own, destroying everything on its path north through the sandy pine country of northwest Wisconsin.

Working the Right and Left Flanks

By 3:00 p.m. Saturday, fire headquarters was full of commotion. People milled around trying to find the sign-up table. Anxious equipment operators asked to get going on the fire after driving long distances. DNR rangers and other employees worked to set up some semblance of organization. All the while more people and equipment were arriving.

Everyone had a sick feeling when they looked to the north. Black, blue, white, and gray smoke billowed up as the flames gobbled different varieties of fuel—trees, dead branches, leaves, and grasses. Nervous energy was abundant, and some people began to criticize the DNR. They wanted to help but were held up in lines, waiting for the sign-in process and for the DNR officials to set up crews, outfit them, find vehicles, and send them out with directions—all necessary steps in project-fire protocol.

Fire boss John DeLaMater seemed like the only one who was calm as he stood next to his vehicle, talking on the radio to rangers Bill Scott, Barry Stanek, and Ed Forrester and participating in calls to and from the Cumberland dispatch center. He knew that the DNR in Madison had just set up a statewide command center. The command center staff was talking to Jim Miller and other fire staff off to the east and organizing manpower and equipment from other parts of the state to fill in behind the depleted forces in the northwest region. Heavy units from Park Falls, Eau Claire, and Brule were headed his way. Ranger stations farther out had started sending trucks and tractor plows from as far away as Tomahawk, Rhinelander, Lake Tomahawk, Trout Lake, and Merrill to fill in behind those coming to the project fire.

The Madison command center was staffed by DNR staff specialist Duane Dupor and chief forester Gordon Landphier. They would spend the better part of three days on the telephones supporting the fire burning in the northwest. It was Landphier who gave the blaze the name Five Mile Tower Forest Fire, because it began near the Five Mile DNR tower. Large fires are commonly given names based on a local land feature or other identifier.

Government weather bureaus were on alert, checking weather maps and providing forecasts to DeLaMater every hour. With the Hudson Bay High pressure system moving over the Midwest, winds could quickly shift from the west—and the entire length of the Five Mile Tower Fire would then become the fire head, blowing and burning toward the east. It would then be not three but thirteen miles wide, racing before the gale, and could, God forbid, turn into another Peshtigo Fire. If that happened, all the bulldozers and tractor plows in the state would not stop it.

His calm exterior notwithstanding, DeLaMater knew he was facing a possible worst-case scenario. He knew forest ecologists and weather experts had done studies on Wisconsin's drought history, superimposing that data on forest-fire records. During the past winter he had listened to presentations and read reports stating that the previous year was the worst drought year in the 116 years covered by the study—and 1977 was even drier. Conditions in northern Wisconsin were as bad as when the 1871 Peshtigo Fire killed more than one thousand people and burned 1.2 million acres. During 1976, DeLaMater knew, Wisconsin DNR personnel had fought 7,821 fires—almost double the normal average annual number of fires. What he couldn't know in April was that 3,923 forest fires would burn in Wisconsin in 1977, for a total of 11,744 fires during the 1976–1977 period.

The most destructive fires occur when relative humidity is lower than 40 percent. Anything lower than 30 percent is unusual for Wisconsin. When the Peshtigo Fire broke out, humidity was 24 percent. On April 30, 1977, the weather service told DeLaMater that humidity was 23 percent. He had quickly decided to not hold back any available assets. In the first fifteen minutes of the fire, DeLaMater had ordered an airplane, the complete overhead team, and traffic control. In the first hour he had summoned fourteen DNR tractor-plow units and ten rangers and assigned one hundred firefighters to various sectors on the fire. All of the overhead team slots were filled in the first two hours, exceeding DNR training requirements.

Now there was no stopping this fire at the head; it was going too fast, and too hot. But the firefighters were holding the east and west flanks. DeLaMater flooded those flanks with people and equipment, exceeding the recommendations of the DNR's Martini Wheel by a factor of three. The methodical task of John Deeres, Caterpillars, and other bulldozers containing the two sides of the fire had begun, foot by foot, yard by yard, quarter mile by quarter mile.

DeLaMater established the first two divisions on the flanks, assigning Ed Forrester the B Division and Barry Stanek the D Division, both on the right flank. Fierce winds were pushing the fire toward the D Division when Stanek arrived with one tractor-plow unit. Dozer operator Pete Paske had made one furrow along this flank earlier, but it was clear it was not going to hold as the hot fire moved through the hardwood slash. Stanek put his tractor-plow unit on this line to double it and sent Northwood High fire crew member Blaine Peterson to follow the tractor on foot, putting out any small fires along the way, in an effort to hold the line. Then Stanek called for more help, asking for two additional tractor plows and hand crews. However, the first unit he received was one of the Minong fire department's tankers. He placed this unit near the intersection of the fire line and a logging road so hoses and water would be available.

Now Stanek saw a breakout of fire near the edge of his division, to the south. He put two hand crews on that flare-up to try to hold it. He put the next John Deere 450 that showed up on this breakout to plow around it and double up the furrows.

The eastern edge of D Division was the Totagatic River. Stanek saw the rapid advance of flames and realized that without massive amounts of equipment, he could not keep pace with the spread of the fire. He ordered more equipment and hand crews, but then, as the fire spurted flames two thousand feet high and smoke columns billowed ten thousand feet up, Stanek's portable radio was failing.

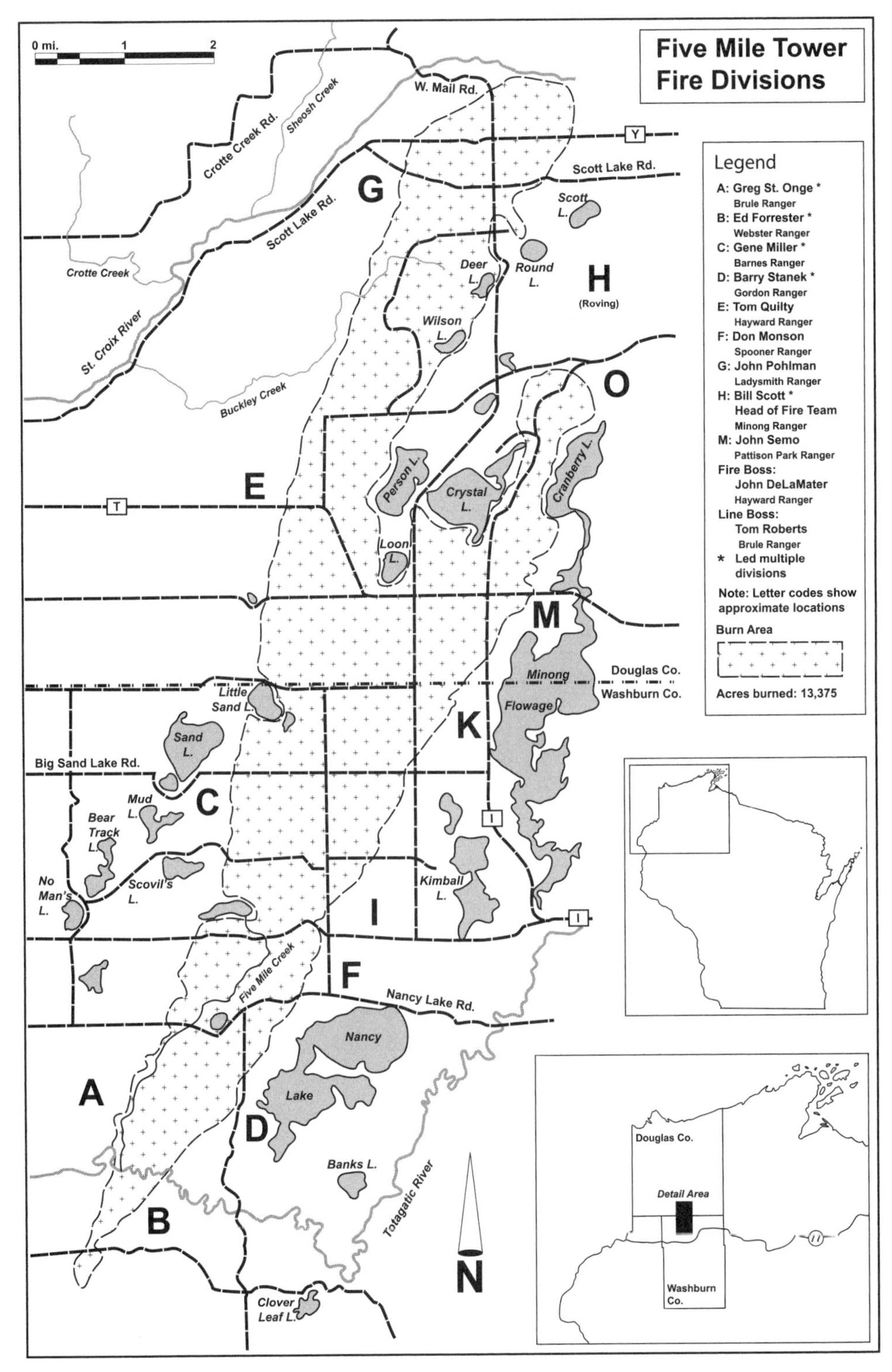

Map by Joel Heiman

Volunteers use their spray cans to extinguish a hot spot near the Totagatic River.

Wisconsin DNR file photo

He had difficulty talking to anyone but the closest plowing units. Of course, he also couldn't talk by radio to Minong fire chief Smokey Smith or any of the Minong fire trucks, because they used different radio systems.

Stanek was struggling to organize his men and equipment to build a double line up the D flank when he heard the fire had jumped the Totagatic River. Two more tractor-plow units arrived, and he continued to build the line. By this time Stanek had four tractor-plow units digging lines, plus approximately ten men walking the furrows, cooling off the hot edge with backpack sprayers and shovels. On the way to the river, one of the DNR tractors caught fire. Had there not been a second tractor nearby to spray it down, it would have been lost.

Now Stanek turned his crews and heavy units around to improve the lanes back south to River Road, their starting point. Stanek called line boss Tom Roberts, informing him he was releasing all units but one and said that the line was holding well and cooling down. With one tractor plow and a hand crew, Stanek would stay until midnight on Division D, mopping up and patrolling the line.

Gene Miller, the ranger from the Barnes station, had traveled more than twenty-five miles to reach the Five Mile Tower fire headquarters by early Saturday afternoon. His station was in the pine region, and

Miller was no stranger to forest fires. He had earned a master's degree in forestry from the University of Minnesota and while in college had many months of fire behavior training in the Boundary Waters Canoe Area Wilderness. He had worked on large prescribed burns as well as large Minnesota forest fires. In 1977 Gene was only thirty but already had five years of ranger experience.

He was assigned to C Division, the second division north of the starting point on the west flank. The fire had jumped the river and a heavy burn was on by the time he arrived. Awaiting equipment, he scouted the west flank for access when tractors and men arrived. He determined that a bridge two miles away was the best route, but getting there required a ten-mile road trip in hilly, tough terrain. Because of the Coriolis effect and the southwestern winds, the fire was trying to turn to the right, and half of the first arriving units were fighting on the right flank. Another 25 percent were assigned to H Division at the moving head of the fire, leaving a quarter of the equipment for rangers on the left flank.

When Gene Miller returned to fire headquarters, he assembled four DNR tractor plows, two private dozers, and two busloads of firefighters. One of these buses was filled with Gordon State Prison Camp firefighters, who brought along their own trailer load of hand tools. It took Miller and his caravan more than an hour to reach the western flank near the confluence of Five Mile Creek and the Totagatic River. The huge trucks with private dozers would need heavier bridges, so routes were established to the west of headquarters, then north, making a much longer trip than was ideal. At one point, the sand road was so steep that each semitractor pulling trailers and dozers had to unload.

The meandering Totagatic zigzagged back and forth. On the ground, it was a terribly difficult area to work. All afternoon and all night, Gene Miller chased the fire. The glow and huge smoke column could be seen far up north in the distance, but his task was to hold the left flank of the fire line. He had to build drivable firebreak lines, sending a DNR tractor plow first, followed by a larger private dozer to push trees and brush out of the way to expand the roadway. Miller and his machine operators and hand crews worked thirteen hours into the daylight of Sunday making new roads that ranger pickup trucks would patrol for weeks, scouting for new fires to knock down.

Greg St. Onge was the ranger at the Brule Ranger Station, poised and ready to fight fire on this dangerous day. When the Five Mile Tower Fire broke out, thirty-one-year-old St. Onge was assigned to travel with his men and equipment to fill in at the vacated Barnes Ranger Station. There was a fire near Barnes that afternoon, burning but contained by dozer and plow. The fire was between two lakes, which made the fight difficult especially on such a hot, dry, windy day. At dusk the fire towers "went down," meaning the tower observers were allowed to climb down and go home to rest. At that time, ranger St. Onge was assigned to travel to the Five Mile Fire.

It was night before St. Onge arrived at headquarters and was assigned to A Division on the western flank, supervising tractor-plow operators and dozers to make drivable lines and burn out the areas between the black and the furrows. He worked with Gene Miller patrolling the lines, both men in pickups carrying water tanks with hoses. The radios were busy, so St. Onge and Miller met face-to-face to coordinate the work of the A and C Divisions. The fire was long gone, racing northward, but leftover ground fires were creeping in the underbrush of the A Division.

St. Onge remained on A Division until daylight Sunday and then was assigned north to Mail Road and the right flank of the fire, approximately three miles north of Highway T. The radios were telling of the possible wind shift to the west-northwest on Sunday. The right flank had to be beefed up and many assets placed on standby in case of a breakout. St. Onge would never forget what happened on Sunday afternoon near Deer Lake, when the dreaded wind shift did occur.

But just now, three huge National Guard D-8 dozers arrived, looking for assignments. St. Onge put them on the north-south fire line near Deer Lake. They roared into action. The monsters on steel tracks effortlessly pushed over trees, digging into the soil to make firebreaks driveable.

Ranger Tom Quilty left his Hayward Ranger Station at 3:09 p.m. and arrived at headquarters forty minutes later. John DeLaMater asked him to report to line boss Tom Roberts. Quilty carried out various assignments at headquarters and finally at 9:00 p.m. was assigned the E Division on the left flank of the fire, just north of C Division. His division extended four miles, from the St. Croix Trail Road north to County Highway T.

Quilty was provided a small army of men and equipment for this long battle, because most divisions weren't four miles long. This was sand country, relatively flat, and filled with red pine plantations and thousands of acres of dense jack pines. There were two lakes on this western border of the fire: Spring Lake on the south boundary of E Division and Little Sand Lake, three miles north. He was assigned a school bus with forty-five firefighters; a DNR John Deere 450 from Grantsburg; four T-6 dozers from Spooner, Prentice, and Hayward; a private John Deere 350 with a front-end loader bucket; a private John Deere 450 with hydraulic blade; plus several pickup trucks. They all drove north to the St. Croix Trail, where Quilty left one John Deere 450 driven by Everett Hanson and Dave Wilson to make a drivable fire lane between St. Croix Trail and Spring Lake. Quilty then moved the remainder of his group up and around to Bear Track Road, where they unloaded and began plowing fire lanes both north toward Big Sand Lake Road and south to meet up with the plowed lane coming up from Spring Lake.

When Quilty reached Bear Track Road, he sent operator Ray Ploof from Prentice south to plow lines back to Spring Lake, along with another private dozer and eleven firefighters. At the same time, he sent the remaining tractor plows north to make fire lanes along with the JD 350 and thirty-three people to secure the line.

All of Quilty's firefighters and heavy units worked, sweated, shoveled, sprayed, and plowed through the night. By 1:00 a.m. Sunday, all of the equipment needed fuel and got filled at both Big Sand Lake Road and Little Sand Lake Road. The hilly terrain near Little Sand Lake had proved to be difficult, especially when the fuel tanks were low on fuel. The engines had trouble maintaining power. It took four hours to plow and secure three miles of fire line. By 4:00 a.m., after seven hours of hard work, the firefighter crew was released for rest.

Quilty also used a National Guard D-7 dozer and a private D-7 dozer to construct fire lanes from Highway T south to Little Sand Lake Road.

Tom Quilty had fought three other fires in the Hayward area on Saturday before arriving at the Five Mile Tower Fire, where he stayed until being released at 11:00 a.m. Monday, a total of fifty-three hours. He drove back to Hayward, showered, and had lunch and three hours of sleep. He was assigned all-night patrol duty on the E Division, beginning at 4:00 p.m. Monday.

The far northwest division of the fire's left flank was assigned to John Pohlman, the Ladysmith ranger responsible for thirty-three and a half towns in Rusk and Barron Counties. He and his Ladysmith crew, driving two Oliver and International Cletrac narrow-gauge dozers pulling sulky plows, had to protect one hundred square miles of Wisconsin's woodlands. In addition to the two plow units, the Ladysmith station had two pickup trucks and two two-ton trucks used to haul the flatbed trailers for the heavy units.

When Pohlman arrived on the Five Mile Tower Fire early that evening, it had already raced forward ten miles. He was assigned the G Division in the northwest quadrant of the fire, a five-mile section from Highway T all the way to the St. Croix River. His duty was to put in drivable fire lanes on the western flank—but this was the area of the fire that contained the massive Buckley Swamp and Buckley Creek areas. He was given a busload of high school students for the hand work, and they were invaluable. As the DNR heavy units and private dozers were plowing furrows down to mineral soil, Pohlman was careful to stay near burned-out black areas where a dozer or group of students could quickly get out of harm's way.

The fire's advance was swift and relentless. It chewed its way north by northeast at the rate of one mile every forty minutes. Only the DNR airplanes and pilots could see the advance in full view. Radio reception was good to average but sometimes difficult. Hills, valleys, rivers, creeks, lakes, sand roads, unmarked bridges, and swamps hindered the firefighting. In two hours, by 3:30 p.m., the fire had raced two miles ahead and had spread one and one-half miles in width.

The Five Mile Tower Fire blazed for sixteen hours.

Wisconsin DNR file photo

An hour later, it was three miles long and one and three-quarters of a mile wide.

All this time, fire boss John DeLaMater was setting up a battle group that by 4:30 p.m. Saturday totaled five hundred untrained firefighters, seven trained rangers, eleven DNR tractor plows, twenty trained overhead team members, three private dozers, and countless back can sprayers and shovels.

Rangers Ed Forrester and Barry Stanek, working with dedicated dozer operators, hand crews, and the Minong Fire Department, kept the wild eastern flank from burning into Nancy Lake, Lower Kimbal Lake, Middle Kimbal Lake, Upper Kimbal Lake, and the southern end of the Minong Flowage. The flames reached dangerously close to cabins and lake homes, coming within one-quarter mile of Nancy Lake. The right flank was a battleground, as crews tried to contain the eastern range of the fire. They worked at a frenzied pace.

Highway T Controversy

By 6:30 p.m. Saturday the fire had burned approximately thirteen square miles of pine forest and was still going strong. It had been burning out of control at the head for five hours. The flame front was almost three miles wide and arching ahead of itself in the winds and searing heat. Most people could not predict that the fire would blast northward for another twelve hours until it would finally be knocked to the ground at the St. Croix River, seven miles distant.

Many who knew the area believed that the fire would be stopped eight miles north, at Highway T. But their company did not include experienced firefighters, foresters, and rangers.

Douglas County Highway T was a major road transecting the southern end of Douglas County. Power lines, bulldozed and mowed ditches, and widened rights-of-way made Highway T look like a wide airplane landing strip. The open swath measured almost three hundred feet wide at places. North of Highway T, dozens of resorts, cabins, lake homes, bars, and restaurants circled Crystal, Cranberry, Person, and Loon Lakes.

Much of the day, the curious drove from both directions along Highway T until they were stopped by roadblocks. There they watched the approaching firestorm. Many had already seen clouds of smoke billowing 10,000 feet into the sky. Imbibers from area bars joined the crowd, some predicting that Highway T would do the job of halting the fire's northward progress. But where were all the Caterpillars? Surely if the DNR would bring up a line of Cats, they could plow another 50 feet of fire line to add to the right-of-way and squelch the fire right in front of them. Nothing could burn across 250 feet of open space plus the blacktop road, they thought.

Bernie Bergman started working on the fire about three miles south of Highway T with a Solon Springs Fire Department tanker truck, saving structures and evacuating people. South of Little Sand Road, the fire was thick and massive in the solid pines. Bergman could feel the wind sucking toward the fire, and for a short while he had a hard time breathing.

Bergman later saw the dozers and their operators along Highway T. They were not authorized to unload their equipment and make a wide firebreak along Highway T. The drivers were parked along the highway, ready to go, with their dozers still on the flatbed trailers. Bergman could see that some of them were angry, griping and complaining, venting their frustration publicly to the crowd that had gathered. Soon the bystanders chimed in. Many people felt the fire was out of control and the DNR didn't know what it was doing. Some who had lost property were crying and angry. Highway T was a major road artery, and there were several bars and restaurants near the roadblocks, so this became a gathering spot from which to watch the smoke and flames. A few observers were emboldened by alcohol and became experts on firefighting, certain they could have done it better.

H Division boss Bill Scott was the local ranger. Everybody knew him, and here he was, organizing and directing the fight at the head of this monster. Rumors on the fire line among the amateur firefighters fueled the hopes that the blaze could be stopped at T. Surely they would be able to all go home in a few hours.

What no one except a handful of experienced DNR rangers realized was that this monster was out of control. Between eight hundred and one thousand citizens were working in the Five Mile Tower Fire by 6:00 p.m. But still it wasn't enough.

A crowd had gathered at Kresch's Harbor bar, restaurant, and store at the beginning of Cranberry Creek, halted there by wardens blocking the roads. There they saw lines of trucks with trailers and dozers, engines running. Most of these dozers belonged to private contractors called to the fire by the DNR. Why, people asked, wasn't this equipment widening the Highway T road right-of-way?

What they didn't know was that recently firefighters in other parts of the country had died when they were placed in the path of crown fires. But the DNR leaders, from Becker at the Spooner headquarters to DeLaMater at the fire headquarters to H Division boss Scott, all knew the risk. They would not allow these contractors, many of them with families at home, to risk their lives in a fruitless effort to try to dig a line in front of this wildfire. And they knew something else: the weather was about to play a horrific hand in this fire. The winds would shift to blow from the west and northwest. The eastern edge of this fire had to hold. If not, a three-mile-wide fire would become a giant, ten or twelve miles wide, heading right for Minong, Gordon, and Solon Springs. There weren't enough pieces of equipment in the entire state to stop a fire that large.

Spooner warden Bill Hoyt hadn't had much sleep of late. The walleyes and muskies were still spawning, and he was working nights with other conservation wardens, trying to catch poachers illegally spearing these fish species, which are helpless in shallow water. On Saturday he was called to the Five Mile Tower Fire headquarters and instructed to begin blocking key roads that would soon be in the fire zone.

He drove up to the H Division and met with Bill Scott. Then he shut down Highway T and got out his camera. As the fire approached, he could see the smoke and three-hundred-foot-high flames coming from the south. It was massive—two, then three miles wide, a rolling conflagration burning at one thousand degrees Fahrenheit.

When a police officer showed up to take over the roadblocking duties, Hoyt drove around several nearby lakes and down back roads, looking for people still in their cabins and homes. Most people living in the pine regions know the danger of wildfire and had already left.

Hoyt had the road blocked on top of the hill above Cranberry Creek near the Kennedy Road intersection with Highway T. The eastern edge of the fire was less than half a mile away. There, at about 6:00 p.m., Hoyt witnessed the most amazing sight of his life. A three-mile-wide fireball rolled forward, sounding like a freight train. Ink-black smoke and white ash billowed through the air. Before the fire reached Highway T, Hoyt looked north and saw hundreds of smaller fires and smoke, caused by embers and flaming tree trunks thrown half a mile ahead of the main fire. He kept snapping pictures, not believing what he was seeing. He thought to himself that there was no way in heaven or hell that this fire would stop at the highway.

Hoyt covered his face against the heat and stood in awe as the fireball rolled across the highway as though it was just a line in the sand. It blew across Highway T, two hundred feet in the air, creating its own wind and spitting flaming shards in all directions, but mostly forward, in front of the twenty-two-mile-per-hour winds.

Trees were already burning on the north side of T when the fire was still half a mile to the south as the wind blew embers and chunks of burning wood forward. The fire crossed over Highway T in a terrifying arch, jumping a half mile ahead of the main body of flames. Dozens of Caterpillars, dozers, and 450s lined up side by side could not have stopped this fire—and most of the men would have died in the heat in such an attempt. You just don't stop a crown fire at its head.

Warden Bill Hoyt's photo of the fire crossing Highway T would end up hanging on the walls of many northwest Wisconsin ranger stations.

Wisconsin DNR file photo by Bill Hoyt

Fire Meets Water

The last ice age had gouged a series of four lakes in these parts: Loon, Person, Crystal, and Cranberry. Now those lakes, like islands in a sea of pines, were directly in the path of the approaching flames. H Division boss Bill Scott had been waiting for this opportunity. Maybe, just maybe, the fire could be slowed and possibly narrowed when the blast furnace reached cool waters. Loon Lake, just a few hundred yards north of Highway T, would be hit first. A half mile north of Loon Lake was Person Lake. To the east lay Crystal Lake, followed by a mile-wide swath of pines and cabins and then the eastern-most, Cranberry Lake.

By this time H Division was short of men and heavy units, but Scott had two men he could count on: Bobby Hoyt from Gordon and George Becherer from Minong. Scott also knew the Minong Fire Department would stand its ground.

About a mile south of Cranberry Lake was the northern end of the Minong Flowage, also called the Totagatic Flowage because the thirteen-mile-long body of water was formed by the power dam that backed up the Totagatic River. View Point Lodge, located directly on the north edge of the flowage and owned by Bob and Dottie Fries, was an important local establishment with a bar, gas pump, and rental cabins. Bill Scott made one of the most critical decisions of the fire when he gave Chief Smokey Smith and ten men from the Minong Fire Department the job of holding the fire along County Highway I and along the driveway to View Point Lodge.

View Point Lodge was surrounded by fire. Owner Bob Fries was on the roof, spraying the shingles with his garden hose. Bill Scott knew he could save both View Point and the lands to the northeast if two

A dramatic photo taken from across the Minong Flowage shows the massive flame front towering above the cabins of View Point Lodge.

Photo courtesy of John Howard Wickland

things happened. They would have to hold the fire along the north-south Highway I blacktop. And they would have to dig a line toward the northeast to the confluence of Cranberry Creek and the Minong Flowage by the Highway T bridge. Here is where trust comes in: the ranger gives the order, and all he can do is trust the skill and bravery of the firefighters.

Scott sent Minong DNR dozer operator Becherer and Webster's Don Crotteau to make a fire line from View Point Lodge north to the south end of Cranberry Creek. Becherer and Crotteau made the line and then doubled back to make another. They did this twice, for a total of four fire ditches, which would improve the chances of pinching the blaze. Scott reminded Smokey that if they let the fire escape to the east, George's and Don's lives would be in danger because they were up there alone with the heavy units, the thousand-degree blaze burning nearby. The fire boiling up from the south was huge, dwarfing the buildings, men, and equipment.

After hours of hard, dangerous labor, the line held. Sometime during the night, View Point Lodge was saved.

The first victory was won. The crews held the fire along the north-south Highway I and up to Cranberry Lake. Just two miles to the northeast were the lakes of Bond, Pickerel, Leader, Whitefish, Warner, Bass, Bluegill, and others, their shores dotted with hundreds of cabins and homes. The devastation of forestland and property would have been catastrophic if not for the deeds of the Minong Fire Department volunteers and George Becherer and Don Crotteau. Bill Scott put them in harm's way, knowing he could count on them to not give up.

After the fire passed over Highway T, it raced into the narrows between Cranberry and Crystal Lakes, threatening cabins on both sides of Crystal Lake Road and along both lakeshores. There, too, many acts of heroism saved cabins. Cabins that did not burn when the crown fire went over the top were saved with sweaty hand labor and strenuous hose work as the ground fires caught up a few minutes later.

Unfortunately, many other structures burned in the intense heat. The fire was moving too fast, even for the Gordon and Minong Fire Departments and volunteers. There, on Crystal Lake Road, is where Tim Foley, another volunteer, and I ran into trouble.

Squeezed into the cab of my Dodge pickup truck with me were Northwood High School teacher Tim Foley and another man, whose name I never learned, who had jumped into my truck an hour earlier, tired from walking on a hand crew. It was about 6:00 p.m. Saturday, and we were about to come closer to danger than we wanted.

We had three backpack water cans and three long-handled shovels in the back and were driving north on Crystal Lake Road, just half a mile from Highway T. Everything was burning on both sides of the twenty-five-five-foot-wide blacktop road. The tops of the jack pines were already dead sticks with no needles, the trunks and twigs blackened but still standing. The thousand-degree heat in the crown blasting forward had seared them like a Christmas tree burned in a brush pile.

Homes and cabins were burning one after another. The smoke was thick, limiting visibility to two hundred feet. Surprised to see one brown cabin still standing fifty feet from Crystal Lake Road, I slammed on the brakes. Fire had surrounded this cabin just a moment earlier. The grass was scorched and all the tree needles gone. A wood pile leaning against one wall was burning. No fire trucks were in sight. Flames trickled up the exterior walls. We grabbed shovels, pitched the burning pieces of oak firewood off to the side, and began shoveling sand onto the walls. To our surprise, it worked. It was more than 120 degrees, and every tree, piece of brush, and blade of grass was either

The fire roars toward a road.

Wisconsin DNR file photo

burned or smoldering. Sweat pouring off us, we kept shoveling until we had extinguished the flames along the perimeter of the cabin and on the cabin's siding. We jumped back into the pickup and drove north into the black smoke with hopes of saving another structure.

We were buoyed with enthusiasm after saving the cabin with nothing but a few gallons of water and lots of sand. We had our windows down as we peered into the woods for more cabins. I drove forward slowly into what looked like a dark cave. I had unknowingly driven into the wild head of the fire. Burning jack pines lined both sides of the road with only a five-foot ditch separating the truck from the trees. We could see nothing ahead of or behind us. In an instant, the hairs on our arms started to singe as flames met in the middle of the road, engulfing the truck. We rolled up the windows, and I backed up. Miraculously, the truck didn't catch on fire.

The noise was tremendous. The roaring wind, cracking bushes, crashing trees, and terrifying howling and moaning made the place into Dante's inferno. I had driven into hell, but I backed out and lived to tell about it.

Less than half an hour after our brush with death on Crystal Lake Road, I drove northward for a mile. The man who had joined us wanted to get out at Wanless Bar, so I stopped and let him out, giving him a shovel and back can. From there Foley and I trailed the crown fire as it leaped from tree to tree in the wind. I spotted a Gordon

Fire Department truck and decided to follow it into the driveway of Earnest and Georgia Madison's lake home on the western shore of Cranberry Lake. Chief Don Finstad and firefighters Bob Williams, Mert Warner, and Craig Smith were there, watching helplessly, as the home had already burst into flames on all sides. It was beyond saving. The outdoor propane tank had ruptured and was shooting flames into the building. The house burned in a matter of minutes.

Roger Finstad, Don's brother, was upset that fire headquarters was more than seventeen driving miles away down Highway 77. The Gordon crews wanted to quickly reach the fire flanks along the lakes west of Gordon, so many just drove out Highway T and were allowed through the roadblocks. Roger joined up with others on the Gordon Volunteer Fire Department headed by his brother. He helped with the crews fighting between Cranberry and Crystal Lakes, establishing a fire line to keep the fire from burning east toward Bond and Whitefish Lakes and the village of Gordon.

Chief Don Finstad directed crews throughout the night, building large fire lines along the east side of the fire. Crews also built fire lines along the west side of the fire along the ridge of the Buckley Swamp and up the St. Croix River ridge. The fire lines on both sides of the fire were about one mile apart and getting narrower. The east and west lines converged at the estate of Russ Hill, where all agencies valiantly fought to save the Hill home.

As the fire burned in the three-quarter-mile-wide narrows between Crystal and Cranberry Lakes, Bill Scott knew he had to make a major effort to build fire lines along the north end of both lakes. He gave this job to Gordon heavy-unit operator Bobby Hoyt, who made fire lines from the north end of Cranberry to the north blacktop road and then south to the north end of Crystal Lake. Here again, one man made the difference. Hoyt plowed into the night, and it was midnight before the looped line was secure. Only a few knew how important this single two-mile, curved fire line would be.

Forester Jim Stordahl's crew also held the line along with Ernie McCumber, Ron Kofal, and a group of firefighters, preventing the flames from crossing the east-west Lost Lake Road. They backfired and, using backpack cans and shovels, made sure the fire was pinched off between Crystal and Cranberry Lakes. It was a critical stand that made the fire half as wide as it was just two and a half miles to the south.

Charles and Einere Marshall had a home and acreage on the west side of Cranberry Lake, in the path of the fire. On his way home from work on Saturday, Charles saw the smoke and fire trucks racing west on Highway 77, but he never dreamed the fire would pass by his home eleven miles to the north. Firefighters staged along Highway T told Charles that he had better evacuate, so when he got home he and Einere quickly loaded up both cars with their three children, their dog and cat, clothing, and valuables, including Einere's new set of pots and pans. An airplane flew over and a voice announced, "Get out! Get out!" The family drove north a few miles to the Five Corners intersection, where they watched the fire dance across the treetops behind them. A little later, the plane flew over again and they heard, "All the buildings along the road were burned."

Thinking their home was lost, the Marshalls traveled to Gordon to spend the night with friends. It was not until Sunday morning when they returned that they learned of the heroics of firefighters who had saved their home and garage, their aboveground gasoline tank, and even the clothes that had been hanging on the line.

They were by no means alone. Martha and George Hinds, owners of Mar Lou's Resort on Cranberry Lake, chose a novel form of evacuation. They took their boat across the lake to wait for the crown fire to blast through. Firefighters saved all their buildings.

The two lakeshores, those plowed furrows, and volunteers with backpack sprayers and shovels saved the night. Bill Scott made sure the lines were burned out; only then would they be secure. The work of Bobby Hoyt and the firefighters not only prevented the fire from burning northeast toward Gordon and Solon Springs. It also cut the fire in half.

Meanwhile, one mile to the west, the fire was split by Loon Lake, on land owned by the Pete Pierce family. Mike Waggoner led a hand crew that cut trees with chainsaws to keep the fire from burning into Pierce's four hundred acres of forestlands.Volunteer fire department hand crews and dozers kept the fire on the east side of Mail Road and

Seventeen-mile-per-hour winds propel the fire forward as it burns through the night.

Wisconsin DNR file photo

pinched the blaze to a stop in the narrows on the south end of Person and Crystal Lakes.

Dale Larson had owned a cabin on Person Lake since the 1950s. Now he was working with the fire crew cutting down trees along the eastern boundary of Loon Lake. He had witnessed some of the conflict between the DNR and private dozer operators and when they had to stay loaded and ready but not utilized near Highway T. But he recalled that most of them changed their tune when they witnessed the terrible blowing and roaring of the fire as it leaped over the highway. Now, by cutting trees along Loon Lake, firefighters saved the Pierce property and forced the fire in a northeasterly direction away from the lake.

Between 8:00 and 9:00 p.m., the fire's eastern boundary was on the west side of Loon and Person Lakes, headed north toward the St. Croix River five miles away. Within four hours the fire shrunk by half. Four lakes and the work of dedicated, tough operators, fearless volunteer firefighters, and concerned citizens had done the job. H Divison chief Bill Scott's plans to use the lakes to slow and narrow the fire had worked.

The last mile-wide flame front rolled across Highway T like huge balls of flaming clouds to the west side of Person Lake. It was there that Bill Scott had a trusted and experienced firefighter whom he knew would make a good crew boss—but he was not a DNR employee. Mike Waggoner worked for the local telephone company. His father was the respected retired Minong ranger John Waggoner, who had fought fires for thirty-one years and had often taken Mike along.

Scott knew Mike Waggoner was up to the task. He issued Waggoner two heavy units and a group of firefighters on foot and sent them to the northwest corner of Person Lake. He also knew Waggoner could backfire with drip torches when necessary. As Waggoner took the men and equipment north toward the St. Croix River, the fire was hot and crowning, racing forward at the rate of one mile every thirty minutes. He and his crew were in the middle of pine plantations in hot, dangerous conditions.

Alex Grymala was at his family's cabin on the east shoreline of Person Lake when he witnessed the fire crossing Highway T west of Person Lake. He watched two fireballs spin through the treetops, three hundred yards in diameter and glowing with sinister energy, hopping over one another with a howling and roaring sound.

As the fire approached, Grymala used his chainsaw to cut down all of the bushes and trees around his cabin and hooked up his gasoline-powered water pump, drawing water out of the lake through two-inch hoses. He wet down the area around the cabin and the cabin itself and then loaded his furniture on a flatbed wagon and pulled it with his farm tractor into the lake. Then he covered his belongings with wet towels and blankets.

Firemen with trucks soon discovered Grymala's water pump, and word went out to trucks nearby. They came for water refills and also discovered Grymala's generosity. In a shed he had a full summer's supply of beer—a stack of beer cases ten feet long by five feet wide and four rows high, covered with wet sheets. Grymala offered beer to anyone coming to fill tanker trucks with water. By the following morning, all of the beer was gone.

Grymala stayed to save his property, help thirsty firefighters with refreshments, and assist many in loading their fire trucks with water. He was a do-it-yourselfer with the right equipment at the right time to help in the firefighting effort.

Near the Grymala cabin was the Blegen family's cabin, also on Person Lake. Kelly Blegen was five years old that summer of 1977; she still remembers the evening unfolding like a horror show as her family watched from their pier. They could smell the smoke when the fire was still a long way off. Then they saw a huge line of fire approaching. A neighbor proclaimed he would not leave his home, saying he would die first. That really scared her. Blegen's parents and Minong fire chief Smokey Smith saved the kids by driving through the roadblocks, past dozers and trucks, to safety on Highway T near the St. Croix River. The fire came within three hundred feet of their cabin, but the fire lines saved it.

Caught in a Fire Tornado

Unusual phenomena occur at the head of a fire, where wind gusts come from all directions, sucked in by the insatiable desire for oxygen. Fingers of flame burn forward, leaving strips of trees and brush untouched. When pine trees explode in the heat, sparks fly upward in a spiral, sending burning shards in all directions to start new fires. The new fires leave green strips that—incredibly—remain unburned in the middle of an inferno.

Another phenomenon that sometimes develops in a crown fire is a horizontal roll vortex: a tornado within the fire. A horizontal roll vortex is an astounding sight—one that few witness and even fewer survive. In certain conditions they become like a fiery tornado lying flat on the ground.

Buck Block and Ed Kofal were no strangers to fire. Both had volunteered on forest fires before, but nothing this big. Ed's brother, Ron, was one of the heavy-unit operators for the Gordon Ranger Station.

A pine explodes from the heat of the fire.

Wisconsin DNR file photo

Block and the Kofals had known each other since they were kids. They had no idea that this night they would experience something few people in the world had ever seen, much less lived through.

When he arrived at the fire, Block was given five Spooner High School students to supervise but no shovels or backpack cans. He took the crew north to Highway T where by this time the crown fire had passed over but the ground fire had not yet caught up. The Minong Fire Department was valiantly trying to save a smoking house. At the same time flames were burning underneath one of the Minong fire trucks and Buck's pickup. Firefighters used hoses to save the trucks and saved the home by dousing it with water.

During the night, Ron Kofal was plowing fire lines north of Cranberry Lake on the west side of the fire, and his brother Ed was acting as his scout. Block and the students were sent to this area of the fire, and finally Ron provided them with backpack can sprayers and shovels. They were somewhere near the Five Corners region at the north end of Cranberry and Crystal Lakes when Ron suggested that Block and his hand crew walk south on a logging road into a swamp to hold the fire line. At this point, the fire was just creeping along the ground. The group had had no drinking water for hours, and when water tanker trucks would come along to fill backpack cans and fire trucks, the boys looked longingly at hundreds of gallons of water they could not drink. Block had to stop one boy from drinking directly from the spout of a water truck that happened to be a sewage pumping truck hired to haul water.

Suddenly, the wind shifted and picked up in intensity, blowing hard from the west. It caught them off guard. The flames immediately leapt high into the tops of twenty-year-old, tinder-dry jack pines. Ed Kofal came running, yelling at Block and the students to get out fast. He had seen the fire blow up some distance away, and it was headed in their direction.

The smoke was thick and the heat intense as the crown fire came over them, surrounding them. This fire tornado was spinning embers and burning twigs, ten feet wide and ten feet high. It was twelve hundred degrees. As wind blew it down to the ground, it produced the rare, deadly horizontal roll vortex. Firefighters don't want to be anywhere close to this phenomenon, much less inside one. Block, Ed, and those five boys found themselves inside a burning tornado. They ran for their lives through the flaming tunnel. Block thought, "This is it!" As he yelled at the boys to run as fast as they could, he tore the spray cans off their backs to lighten their loads.

They ran with fire above them and on both sides. It was difficult to breathe, impossible to see. But then an astonishing effect of the fire saved their lives. The raging fire tornado sucked up fresh air as it swirled, and the boys and men sprinted through the tube of terrifying flames, breathing fresh, cool air and running on unburned grass to safety. All of them made it to the truck. Not a hair on their heads or any of their clothes had caught fire. It was a miracle.

Then Block and Ed saw a private dozer operator trying to drive his equipment out of the blazing flame front. It was so hot he had his coat over his head and had put on gloves. Ed and Block ran to help him, but the driver did not see them in his haste to leave the area. The man driving the dozer had one thing in mind and that was to make it to his truck and flatbed trailer in one piece. He barreled up onto the trailer at full speed, almost running over Ed and Block, left the dozer running, and drove off in a spray of sand and gravel. The two men had never seen a dozer operator so scared before—or one so close to death.

Then, just as suddenly as it had shifted, the wind dropped in intensity and returned its northeastern flow. The horrifying incident was over. It was 1:00 a.m., and Block and his crew had not had any water or food in seven hours. Buck piled the exhausted boys into his truck and took them to his home in Minong to give them water. Then he drove the boys to each of their homes, twenty miles south in Spooner.

Later the next day, Block took his brother Jim to see the exact location of the fiery tornado. They looked at the unburned strip of forest and grasses and found the discarded backpack cans. Amazed, they gazed down at the burned five-gallon metal spray cans lying there on the ground. The wooden handles were burned off, and all of the rubber gaskets and rubber hoses were melted. The cans were now just scorched piles of aluminum and brass fittings.

The Breakout

Everyone warned about a breakout. The fire lines were holding, but everything inside was hot and still smoldering. Brush piles and stumps still had fire deep inside them, down into the roots. A sudden wind could give enough oxygen to one of these sleeping hot spots to cause it to blow up into another ground or crown fire—called a breakout. It was the very thing DNR rangers did not want. The weather reports were ominous, predicting wind velocities and directions that might cause the entire eastern flank of the fire—thirteen miles long—to suddenly become the new head. All night long, the bosses at headquarters sent more and more equipment to the east flank to widen the fire lines. They were afraid that if the wind did shift, they would need lots of luck plus a wide firebreak to have a chance of containing the blaze.

The fire line between Person Lake and the St. Croix River, a distance of four miles, was black and dug into mineral soil. The line was holding, but headquarters sent additional dozers anyway.

A dozer operator from Brule, Buck Follis, was driving one of the very few new John Deere 450s on the fire. Earlier on Saturday, he had fought a fire in Barnes on Sandbar Lake. When he was sent to the Five Mile Tower Fire, his route took him along Highway T, where he saw the fire leap across the road. He unloaded his heavy unit and began plowing near Crystal Lake all night, finishing at the Wascott Dump at daylight.

By midday, after working for thirty hours nonstop, Follis loaded his dozer onto the flatbed trailer and went to sleep in a grassy ditch.

Soon the word came over his radio: breakout! Every unit nearby was ordered to Section 1, just south of Scott Road.

Ranger Greg St. Onge was patrolling the fire lines near Scott and Daisy Roads and saw it happen. Even though the lines were wide, a pile of brush might have contained an ember, or a stump might have been cooking inside for hours and then suddenly erupted into flames. The breakout was burning slowly, ten to fifteen feet across, as St. Onge approached with his four-by-four and hose reel to spray water on the burning pile of branches. Suddenly the wind picked up, blowing from the west as predicted. As he was spraying water from the one-inch-diameter hose, the wind caused the flames to instantly jump into the jack pine trees nearby. One second the fire was on the ground, and the next it was twenty feet up, roaring in the treetops.

The fire ran away from him in a southerly direction into unburned trees and brush. St. Onge called on the radio for help. Follis was soon there, plowing on a fire that was now going hot and fast. George Becherer also arrived with his narrow-track dozer and steel-wheeled plow, digging a fire line around the breakout. It was a frantic few minutes as the men worked to prevent the breakout from marching to Gordon. The breakout was about one-third-mile long and one hundred yards wide, up in the trees and racing red hot and furious. Becherer hooked an oak tree with one wheel of his plow, tearing the steel and breaking off the wheel. He kept going. The dozer operators could see one another on each flank of the flames, because the breakout was narrow.

The fire was prancing faster than a person could run. Chunks of flaming debris spread the flames forward, prompting St. Onge to call Bill Scott for additional help. Soon two large D-7 National Guard dozers showed up. Then came a hand crew with shovels. St. Onge told them to follow the fire ditches and throw sand on anything burning.

Equipment and crews held the breakout in check. Then, just as suddenly as it had whipped up, the wind dropped in speed. The fire dropped out of the trees and back onto the ground, allowing the dozers and firefighters to encircle the area and halt its march.

Happiness reigned when the firefighters saw the smoke column disappear and the flames get beaten down.

Fleeing the Flames

My wife, Karen, got the phone call Saturday night. The man on the phone told her to evacuate our home on the north side of Bond Lake, just three miles from the eastern flank of the Five Mile Tower Fire. The fire bosses were expecting a wind shift from the northwest, and they were scared.

Karen packed our Volkswagen Beetle with valuables and important papers and got our two young boys—Mark, age three, and Greg, six months old—out of bed. She hoisted our two dogs and cat into the car. When she drove out of the driveway and turned south on the blacktop, she was horrified at the sight. In the west the sky was red from horizon to horizon. Flames from three miles away could be seen licking toward the heavens. By the time she drove out of harm's way toward Minong, the fire had scorched ten miles of forest and still had another four to go.

Hundreds of people were evacuated that day and night as the flames raced northward. Wardens, police officers, and firefighters went ahead of the fire's anticipated path and notified folks to leave. Because of the forecasted wind shift, the evacuation plan included lands up to five miles east of the fire.

Marilyn Scott was home with her young children, two miles west of Highway 53 at Scott's Resort. Her husband, Dick, was busy delivering sandwich meat, cheese, and bread to the Sandwich Queens of Minong and then taking the sandwiches to the fire zones. As the fire grew in intensity, Dick stopped at home to check on his family. They loaded their car with their children and some belongings and went to the Wascott Town Hall, which was opened for evacuees. Dick then returned to his volunteer work, delivering food all night long.

Ed and Deb Kofal lived along Highway 53, next to Halfway Lake, between Wascott and Gordon. Deb and her sister-in-law, Cheryl Smith, watched the fire grow and decided to evacuate. Deb loaded important papers into their car, along with all of Ed's carpenter tools. The two women went to Minong to stay with Deb's parents that night. When Ed returned home at 7:00 Sunday morning, he thought someone had robbed the place.

Jack and Mary Morrow's house on Bear Track Lake was on the western edge of the flames when they were told to leave. Earlier in the afternoon Jack and their son, John, two and a half, were fishing under the Totagatic River Bridge on Highway 77 west of Minong when they heard the first DNR trucks and heavy units going overhead. They climbed the riverbank and saw the first columns of smoke in the distance. Little did they realize that their home five miles to the north would later be threatened.

The Morrows evacuated to Ed and Betty Nelson's home in Minong. Jack's mother evacuated from her home on No Man's Lake. Meanwhile, Jack and his father, Jack Senior, returned to their homes and stayed on watch with garden hoses ready all night. Jack Senior opened the lock on his gasoline tank and gave fuel to any trucks in the area needing a fill-up.

Jan Mednick was so nervous trying to get out of the way of the fire that she set her cat litter box on the driveway, put her cat in the car, and in her haste got in, started the car, and backed over the cat's commode. Jan was grateful that Eugene Barrett, the Minong Township chairman, came along in his car while evacuating residents and offered to lead Jan to town on roads that were not blocked by fire vehicles. She followed him down barely passable sand roads dug up by trucks and dozers, through the smoke and haze and out of danger.

Kresch's Harbor store, tavern, and restaurant, located at the Highway T intersection with the Minong Flowage, had become a gathering spot for many who evacuated from the fire. They parked their cars and trucks, drank beer, and watched the towering flames just one mile away. Most had opinions as to how the DNR was doing in fighting the racing crown fire. Dottie Fries, owner of View Point Lodge, was there, too. She overheard discussions about what valuables people saved from their homes before fleeing the flames. One lady had rescued her three small pigs and was hauling them in her car trunk. Another woman could think only of saving her large box of hair curlers.

Citizens Fight the Fire

Thousands of stories linger in the minds of those local residents who lived through the fire. Some are sad, many are touching, and a few are even comical.

Michael Edwards recalls watching the orange glow in the sky even though he was only ten years old at the time. He lived twenty miles north of the flames and remembers the strong winds that evening. Later, he and his father walked over parts of the burned area and were surprised at the damage. They came upon dead wildlife everywhere. Now, Edwards is a wildlife specialist for the U.S. Department of Agriculture and has worked on the restoration of the Five Mile Creek near Minong. He witnessed the regrowth and rejuvenation of the woodlands after the fire.

David Babcock and his friend Jerry, resort owners, signed up as volunteers on the fire at headquarters and were told, "Get on that bus." After sunset, after fighting the fire for hour upon hour near the head of the flames, they were exhausted and cold. While they were trying to get warm around a still-burning stump, the crew boss yelled at them to put out the flames and get back to work at the head of the fire. They yelled back, "Hold on, we are just getting warm." The boss insisted they "get a move on!" Babcock yelled back, "Listen, we are just volunteers. We just want to get warm for a while." The crew leader ran over and said, "Volunteers? This is the Gordon Prison Camp crew! You got on the wrong bus!" They fought the fire with the prisoners until daylight and finally sat down with them and had hot coffee and sandwiches before going home to rest.

Seventeen-year-old Pat Conley's grandmother owned a cabin on Little Sand Lake, on the western edge of the fire. Because they had heard all of the roads were blocked, Pat and his girlfriend, Deanne, asked a pilot friend, Oscar Amorde, who lived in the town of Oakland, to take them flying over the burned areas to see if the cabin was still intact. They rented a plane at the Superior Airport and flew in turbulent conditions over the fire. Pat had to hold an airsickness bag for Deanne during the flight. They were married six years later. Deanne said that anyone who would hold an airsickness bag for her must be a pretty good guy.

Spooner High School student Ricky Black, his brother, David Barrett, and friend Brian Kamin all were driven to the fire lines in a van and were strong enough to carry the heavy backpack can sprayers. They worked for days straight, stopping only for catnaps in ditches and occasionally going back to staging areas for food. Twice they stopped beer trucks on the road and were given cases of soda. Thus fortified, they went on to save several cabins.

Jeff Kline, a Northwood student, was digging fire lines with shovels along with a group of students when they were picked up and driven to the Buckley Swamp, where there was a flare-up. Heavy equipment could not get at it, so they hiked in single file following a DNR ranger. The fire was on a small island bordered by a creek that looked about two inches deep. As they marched across a fallen log, the ranger fell off the log and quickly sank up to his head in muddy peat. Only his hat was dry. The boys helped pull him out of the quicksand-like mud, and the group finally got to the flames. The creek was so muddy that the back cans kept plugging up from the peat, so they put the burning area out with shovels.

Mark Smith, a Minong resident and teacher for the Grantsburg School District, sustained the only injury on the fire. He was working with the Minong Fire Department, standing on the rear platform of a fire truck and holding onto the safety bar. They were near Three Mile Road when the fire crossed the sand road both in front of and behind the trucks. They were trapped between two masses of flames and hot gases. The truck driver started backing up away from the fire in front, but he did not know he was backing right into flames behind them. Smith got so hot that he jumped toward the ditch and rolled away from the rear wheels, spraining his ankle. The flame front went past, and the trucks and men were saved as Smith limped back toward the truck and got on board.

Brian Huntowski was too young to fight the fire, but he recalls hearing on the radio about the need for volunteers. As he rode along with his sister, father, and older brother as they drove to the fire on back roads, Huntowski saw burning animals running out of the fire—rabbits, squirrels, and deer. He saw wardens and other workers hit suffering small animals with shovels to put them out of misery. The smells and heat were frightening, and when his father and brother went to sign up and get hand tools, he was worried that he would never see them again. At home later, Huntowski and his mother worried all night and got little sleep. The sun never seemed to set that night, he recalls. The next day, the smell of smoke lingered in the air. His dad, Pete, and brother, Tom, finally returned home, black from ash and exhausted. His father told of a tree exploding right next to him and almost melting his glasses. At another point on that long Saturday, Pete and Tom were forced by approaching fire to crawl into a lake. It felt like hot bathwater.

Tim Foley fought the fire all weekend, saving cabins with hand tools and witnessing the brunt of the flaming front as the fire crossed Highway T. In the Buckley Swamp, Foley was carrying an aluminum back can sprayer during the mop-up operations, and someone sawing down burning trees dropped one on Foley's back. The falling tree didn't injure the six foot three, 215-pound Foley, but it crushed the spray can. Foley continued working on the fire on Monday, but in an official capacity, as a teacher–crew leader for Northwood High School.

Northwood High School math teacher and part-time cement contractor Jim Block was working with his brother Buck that Saturday, building a basement block wall in a Minong neighborhood, not far from the home of DNR ranger Bill Scott. As Jim and Buck worked on that hot, windy afternoon, they could see the smoke column glowering above distant trees. They also observed the Scott home becoming a beehive of activity as neighbors gathered there to make sandwiches for the fire workers.

When they finished the wall, the Blocks went straight to fire headquarters to sign in, still in jeans and T-shirts. The brothers were separated, Buck heading to the flaming north end of the fire and Jim assigned to a crew working in the blackened burned areas to the south. Jim's small group worked all night turning over burning stumps and putting out small fires in brush piles. When the sun went down the temperature quickly fell, and Jim and the other volunteers in his group were freezing cold because they had no coats or sweatshirts.

Those hundreds of sandwiches being made back in Minong were a distant memory. For it would not be until 3:00 a.m.—after eleven hours of work—that someone arrived to provide Jim and his crew some beverages and sandwiches.

It was 2:00 a.m. Sunday, and I had just helped save the Hill home. I was worried about my wife, Karen, and sons, Mark and Greg. I knew there was a pay phone on Highway Y across the street from the Lookout Tavern near the Gordon Dam. I called my home, but there was no answer, so I dialed my parents, Gladys and Matt Matthias in Madison. I woke them and asked them if they knew where Karen and the boys had gone. I knew the fire had not burned as far east as Bond Lake. They told me that Karen had called them and had evacuated to the home of one of our friends in Minong. While I was on the phone with my parents I described the scene I saw from the hill high above the Gordon Flowage: the glowing sky, the embers shooting up hundreds of feet, and the smoke rising toward the stratosphere. It looked like the entire horizon was on fire.

Tom Brisky and some buddies were shooting pool at View Point Lodge Saturday afternoon when someone came into the bar telling about the fire. The group left in their pickup trucks, and Tom joked that he hoped the fire would not reach his new home on Three Mile Road. The house was six miles from the origin of the fire, and no one dreamed the fire would get that far. Officials at fire headquarters on Highway 77 sent them to hold a fire line, putting out embers and burning logs and stumps, in an area where the fire had already done its damage.

After four or five hours, Brisky's crew was loaded up and sent north, to near his home. The area was devastated. Miraculously, Brisky's home was spared, although his neighbor's cabin had burned to the ground.

He fought the blaze until daylight and then drove to his house, got a ladder, and climbed onto his roof. He wasn't prepared for what he was about to see. There wasn't a tree in sight; everything was black. He wondered how long it would take for regeneration.

The Final Battle

Mert Warner, who had served across Europe as a medic in World War II, battled the Five Mile Tower Fire to the ground.

Warner was working with other members of the Gordon Fire Department, including Bob Williams, Don Finstad, Craig Smith, and Steve Balcsik Sr., when the fire passed right over them. Although Balcsik was so scared that he drove the fire truck away from the scene with hoses dragging behind, they had been spared to go on saving homes and cabins along the road.

All night they followed the dozers on the hills to the south of the St. Croix River with backpack cans and shovels. Dozers and hand crews finally extinguished the fire near the Russell Hill estate and Highway Y, the site of what is now Fire Hill Golf Course, at the north end of twenty-two square miles of burning and smoking rubble.

It was in the early-morning hours of Sunday and just getting light. High above the St. Croix River, the firefighters put lines around the flames, knocking them to the ground on the Hill property. Barry Stanek was there, just seven miles from his Gordon Ranger Station, supervising the last efforts.

The wind had dropped. The fire had narrowed to less than three-quarters of a mile at the river, fifteen miles from where it started. A DNR fishery crew was working in the river with backpack cans, shovels, and portable pumps. Stan Johannes, the DNR fish manager from Spooner, was there with his men, racing back and forth across the shallow river and over slippery rocks, putting out fires as licking flames crept down the hillsides. Flames blew across the narrow river, but workers quickly extinguished the flare-ups with water and shovels.

It was there that the Five Mile Tower Fire ended as it had started, a small, flickering flame.

Extreme fire-danger conditions had contributed to the Five Mile Tower Fire's intensity and size. Finally, on Sunday afternoon, conditions of a different sort helped put it out. The dogged efforts of sixteen hundred firefighters during the previous twenty-six hours had pinched the fire almost to a point, less than a mile across. The dangerous Sunday breakout was contained quickly, and the racing crown fire was knocked to the ground. On the slopes near the St. Croix River, the vegetation changed to more hardwoods and river-bottom growth, and the soil changed from dry, sandy pine plantations and dense jack pine growth to heavier, wetter soil.

But the most important change was the wind. The howling southern winds quit during the early-morning hours on Sunday. With this drop in wind velocity, the flames moved to the ground where workers could more easily contain them. Many dozers with plows were there at the end, along with hundreds of men and women with backpack cans of water spraying and shovels digging—firefighters who now were experienced, who had been battling these flames since the day before.

Crews working in the river both on foot and from flat-bottomed boats with hand tools and portable gasoline-powered pumps with hoses put out the last of the flames—some as they crept down the southern slope and some that had jumped across to the opposite northern shoreline of the river.

Most of the firefighters were not there to see the last flames extinguished. More than a thousand had already headed home by noon on Sunday, and the five or six hundred remaining were spread out over the twenty-two square miles of blackened earth, many assigned to patrolling for burning stumps and logs or for smoke. Most of those who remained on the scene had not slept, and some were delirious from exhaustion and lack of food and water. For some, the fire's end was anticlimactic, as they looked up with bloodshot eyes and soot-streaked faces at a sky now devoid of the horrendous billowing smoke columns. Word got around that the fire was knocked to the earth up by the St. Croix River. There were no joyous celebrations. No one jumped with happiness. Folks were too tired to jump.

H Division boss Bill Scott had battled the fire all the way. Now, there was nothing behind him for fifteen miles but destroyed forest. On Sunday afternoon, after the breakout was put down and the fire burned into the St. Croix River, Scott drove south for the first time in twenty-six hours. The winds had died, and things were calm. He drove past the smoking wrecks of buildings and smoldering trees, crossed Highway T, and drove down a sand road to an intersection at the center of where the fire burned. He got up on top of his truck and was chilled at the sight. He almost did not know where he was. It was like a moonscape—unrecognizable. The entire landscape was gray, black, and brown as far as the eye could see.

Mop-Up

The main fire was out, but its wide path was still hot, smoking, and smoldering. On Sunday afternoon, May 1, after the final flare-up, the crown fire was down to the ground and contained with forty-three miles of bulldozed, drivable sand roads. This ribbon of brown sand separated the green trees and brush outside of the fire zone from a blackened interior of 13,375 acres. There was no rain over that terrible weekend and none in the forecast, and fire boss John DeLaMater, who had not slept since Friday night, needed help.

Sunday afternoon, hundreds of firefighters were released to go home and sleep. By that night, hundreds more had left the fire scene. Men and women who had battled so hard on the fire had to return to work on Monday morning. Hardly anyone was left to do the ugly task of mop-up. Burning stumps, smoking brush piles, logs hot to the touch and glowing on the inside, and standing timber, blackened but still smoldering, all needed to be put out by hand labor.

I was at fire headquarters late that afternoon when DeLaMater asked for the help of the school students. I saw the rangers with bags under their sunken and distant eyes, dusty, unwashed, sunburned faces, and wrinkled clothing. Many of these men had not rested since Friday night. Some had actually been at it since Thursday night, fighting other fires. Still, the managers of the Five Mile Tower Fire were not planning on sleeping Sunday night, either. The fire zone was too hot to relax. Rangers and heavy-unit operators would have to keep twenty-four-hour-a-day vigils until it rained. They went on twelve-hour shifts—and were expected to take care of their own ranger station areas at the same time. They needed help, and I pledged to provide some.

I went home and called Northwood High School teachers whom I knew would be willing to be crew leaders for the students. When I knew I had the supervision necessary for the task, I began to telephone the TV stations in Duluth and the radio stations used for winter-weather school dismissals. Spooner, Shell Lake, Hayward, and Duluth radio stations informed the Northwood students to come to school on Monday morning dressed for outdoor work if they wanted to help. The work was voluntary, and kids who did not want to go were assigned study halls or could stay home. I asked school cooks to come in early and prepare sandwiches and fill cases of chocolate and white milk to take along on the buses. As superintendent, I later heard criticism from some people for the school's support of the fire-suppression efforts, but I knew we had done the right thing in a time of community catastrophe.

On Monday morning we were ready. After the school buses completed the morning routes, they were gassed up and parked in front of the school ready to transport crews of ten students and one teacher crew leader to fire headquarters for sign-in. As soon as the students left the school building they became employees of the DNR, paid $2.30 an hour and covered by workers' compensation insurance. School secretaries Cheryl Smith and Jean Legg kept accurate logs of the names and addresses of every student and teacher going to the fire. Cooperation was the name of the game. Many of these same students had had little sleep all weekend because they were on the fire. Young men and women now stepped up to the remaining task—putting out the final smokes of the largest forest fire in Wisconsin since 1959.

Student members of the primary and secondary fire crews at Northwood High School were asked to obtain permission from their parents to stay out all night on the fire lines if necessary. The student fire crews had ten members each, but most of the high school student body participated in the fire mop-up operation.

In addition to the Northwood students doing the grunt work following the fire, students from Solon Springs, Hayward, and Spooner High Schools participated. Principal Darryl Schnell of Spooner High School allowed many students to travel to Minong and work on Monday, Tuesday, and Wednesday. Fire boss DeLaMater assigned rangers and other DNR employees to take each group of students to different locations. In addition to having teachers along, the students were supervised by fire specialists.

Northwood senior Bonnie Burnosky put on many miles mopping up the fire along with her classmates. She weighed all of ninety-five pounds and carried a fifty-pound backpack can. Years later she still recalled that she had worn her favorite knee-high moccasins but ruined them in the ashes, dirt, and burning embers.

One of the Spooner students working on the fire was Bill Thornley, who also had a summer job with the *Spooner Advocate*, the first newspaper to print stories about the Five Mile Tower Fire. Thirty-three years later, Thornley is the editor of the *Advocate*.

As the students worked in the burned areas, they were amazed at all the buildings still standing amid the burned forest. They worked in the vicinity of many homes, cabins, and garages that made it, thanks to the bravery of the firefighters over the weekend.

These happy sights were in stark contrast to the animal skeletons lying about and the blackened piles of burned lumber, scorched bricks, and bare concrete sidewalks leading to the ruins of five trailers, four homes, fourteen cabins, six garages, twenty-three storage buildings, four barns, and seven sheds.

On Wednesday of mop-up week, teachers Jim Block and Doug Denninger were with crews of Northwood students in Buckley Swamp when an ember smoldering inside a tree blew out and started a new fire. This was in the far northwest part of the fire zone, and a DNR airplane pilot had to tell ranger Ed Forrester the direction to walk to the burning area. DNR workers came to help, but George Becherer's narrow-gauge dozer tipped over in the soft peat, and Ed Forrester fell into a sinkhole so that all that was showing was his hard hat. The group had to get down on their hands and knees to breathe under the smoke. The students put out that fire with water and shovels and, without the help of Becherer's dozer, lots of hand labor.

On Thursday and Friday, the Northwood students finally returned to school. Classes were held, and the gym was nicely decorated for the prom that weekend.

The students were eager to receive their checks from the DNR, but they also knew they had just done something special. For three days they had been part of citizenship in action. They worked hard to protect the entire community, put that fire out, and clean up. They had learned real-life lessons in volunteering in times of trouble.

During that week of mop-up work, the nights were stressful for the tired rangers, who were patrolling in twelve-hour shifts. There was no rain. Six arson fires were started during the first week of May along the perimeter of the Five Mile Tower Fire, requiring the dispatching of nearby dozers and trucks. John DeLaMater even sent up DNR airplanes to try to spot the arsonists. No one was caught and little burned, but it was just one more difficulty the rangers had to endure that spring.

Young Bud Schaefer continued working fire standby on weekends for ranger Bill Scott in Minong. Two weeks after the fire, Scott asked Schaefer to hike into a swamp along Five Mile Road. There Bud found some smoldering stumps, which he put out with the backpack can. They patrolled the entire perimeter and found no more smoke. On the way home that day, Scott bought Schaefer a hamburger at Sleepy Eye Resort and said, "That may have been the last 'smoke' of the Five Mile Fire!"

On Saturday, May 13, two weeks after that infamous campfire started the Five Mile Tower Fire, Scott knew it was extinguished. Fire boss John DeLaMater agreed. He called the DNR in Madison and officially declared the fire was out.

Aftermath

One match sparked Wisconsin's largest single-source forest fire in fifty years.

The Five Mile Tower Fire raged without stopping for seventeen hours. It was a crown fire, long and narrow: fifteen miles long and three miles wide at the widest. For people living in this region, it seemed as though their whole world was ablaze.

During the afternoon and evening hours of April 30, 1977, more than four hundred people were evacuated by conservation wardens, county sheriff deputies, and the public-address speakers mounted on the bottom of DNR airplanes. Forty roads were blocked to keep onlookers safe from the flames and potential looters away from the abandoned cabins and homes.

As the fire advanced, steadily chewing up one mile every forty-one minutes, nine volunteer fire departments came to assist in working the fire lines along the edges. Their knowledge, speed, and bravery also saved a total of 323 buildings. Sixty-three structures were lost, including some people's only homes.

Helping the fourteen DNR dozer-plow units were thirty-one private bulldozers. The National Guard provided six huge dozers, operated by the Wisconsin National Guard 724th Engineer Battalion.

In total, forty-three miles of drivable twenty-foot-wide fire lines were plowed around the fire perimeter. This is the equivalent of the highway department cutting a rough road from Gordon to Superior, or from Madison to Janesville—through the middle of a forest, over swamps, around lakes, and over rivers—in less than one day!

The first eight miles of the fire burned mostly Mosinee Paper Mills forestlands, and the final seven miles blackened primarily Douglas County forestlands. Some private plantations were lost as well.

In all, twenty-two square miles burned.

The tragedy and the sheer immensity of the fire galvanized the entire regional community of Spooner, Minong, Wascott, Gordon, and Solon Springs. About sixteen hundred people fought on the fire, but that number doesn't include all of the police officers and ambulance emergency medical service personnel on the scene in standby mode, the hundreds of women and men making food, or all of the men and women driving water tanker trucks of all shapes and sizes. It doesn't cover individuals delivering diesel fuel or gasoline to the fire zone or the many private dozer operators and National Guard dozer drivers. In the end, probably well more than two thousand people worked to put out this menace over a two-week period.

Most amazing is the fact that no one was killed, no one was seriously injured, and there were no reported hospitalizations as so many people worked in such dangerous conditions. Many came close to danger, and some even felt the flames up close, but everyone went home to their families, school friends, and relatives.

John Schultz, the landowner who unwittingly unleashed an inferno when he lit a match to start a campfire, would be charged by Washburn County district attorney Patrick H. Stiehm with a misdemeanor violation of statute 26.14(6) for starting a fire and allowing it to escape. Schultz stood trial in March 1978 in Washburn County Court before Judge Warren Winton. Other than a brief newspaper account by the *Spooner Advocate* and the memories of those in attendance, there are no records of Schultz's trial; according to the Washburn County district attorney's office and the Washburn County clerk of courts, records of misdemeanor trials are kept for only a short period. At the time of the Five Mile Tower Fire there had been no burning ban in effect, and the jury felt that Schultz had taken reasonable precautions in preparing the fire pit and trying to keep the cooking fire small. On March 31, 1978, the jury found Schultz not guilty.

When the jury's decision was announced, the prosecuting district attorney Stiehm stepped over and gave Schultz a congratulatory hug. Several of the DNR rangers in attendance who had worked hundreds of hours on the fire quite naturally cringed.

At the time of the Five Mile Tower Fire, Mosinee Paper Industrial Forests (now Wausau Paper Company) owned 75,000 acres in northwest Wisconsin. As the fire raced through Washburn County during the first five hours, most of the land destroyed was owned by Mosinee. In all, the company lost 4,426 acres. It was a major blow, and it would be another thirty to forty years before any pulp wood could be taken from the devastated seven square miles.

Terry Michal, Mosinee Paper's lead forester, and Steve Coffin, second in command, were called to fire headquarters with their ten-man logging crew to help with the fire and offer assistance in intelligence and mapping. The Mosinee men were assigned to the hot east flank of the fire after it had jumped the Totagatic River; there they had hooked up with Bobby Hoyt's DNR dozer and plow from Gordon and worked on the dangerous right-hand divisions for many hours. These were experienced woodsmen and were given the toughest assignments. At 3:00 a.m. they were transferred all the way to the St. Croix River to help with the firefighting in the northeast sectors. After more than twenty-four hours on the fire, the Mosinee crew was released.

The Mosinee employees' fear that their losses would be catastrophic were confirmed by later timber cruising and airplane observation mappings. A single landowner lost 32 percent of the total acreage burned in the Five Mile Fire. That was the Mosinee Paper Industrial Forest.

In the fire's aftermath, what was the company to do with all of the burned timber? The paper mill couldn't use the charred trees, because pieces of black carbon embedded in the standing dead timber would show up in the pulp and later in paper products. The company tried a novel approach and sold the logs by the millions to chipping operations. The entire tree—trunk, branches, cones, and all—went into the hopper for pulverizing. The wood chips containing black ash particles and blackened bark were sold to turkey farmers for turkey bedding. The company got rid of the burned wood, and the turkey

The Five Mile Tower Fire rages behind a grassy former farmer's field dotted with jack pines.

Wisconsin DNR file photo

producers obtained nice pine bedding with charcoal pieces, useful for smothering smells, much as it does when utilized in filters for air masks and fish tanks.

Still, the impact of the fire losses gave Mosinee Paper a renewed awareness of its responsibilities in fire protection. The company purchased a new dozer and attached a fire plow to it and also bought slip-on tanks and hoses for the pickup trucks.

In 1980, three years after the Five Mile Tower Fire, the Oak Lake Fire started just a few miles down the road. As it burned south, pushed by a hot north wind, it destroyed 11,418 acres. It was then that the foresters got worried. Pine covers northwest Wisconsin in huge blocks from Grantsburg northeast all the way to Iron River. The fire destruction and inability to stop a crown fire in the pines led the foresters to consider massive firebreaks as a part of forestry thinning.

Mosinee Paper, along with several county forestry departments, clear-cut quarter-mile-wide swaths of timber for pulp and sawed-wood production to create a firebreak that would have a chance of containing a large fire. These corridors were six or seven miles long. Several years later, when the cleared strips started growing young trees again, the foresters would go north or south and clear another long path.

Russell Hill owned almost two hundred acres along the St. Croix River. At the time of the Five Mile Tower Fire, the U.S. Park Service had been trying to purchase Hill's property under the Wild and Scenic

Rivers Act with plans to remove all of his buildings and return the land to pristine wilderness. Hill had built many buildings—garages, painting studios, storage sheds, and, in the midst of the forest, an authentic western village, complete with Chinese laundry, Wells Fargo station, barber shop, sheriff's office, general store, and blacksmith shop, all purchased out West, hauled to Gordon, and reassembled. The buildings contained priceless antiques and equipment. Hill had steadfastly refused to sell.

As the fire approached his estate, Russell Hill was seen in his pickup truck several miles away on West Mail Road. Gordon Fire Department members yelled at Hill to get back home; his place was threatened and his wife was there, in danger.

When the fire raced through his acreage and burned many of his buildings, Hill was devastated. Of all his buildings, only his home was saved—by heroic effort. He thanked the firemen and the DNR.

After a fire, regeneration and renewal come slowly. Every year on the anniversary of the Five Mile Tower Fire, Minong ranger Bill Scott would drive to the same intersection of Highway T and the sand road, the epicenter of the fire. There he witnessed the changes. The first year there was some greening of the ground plants, and many of the standing black trees were gone, harvested for grinding and sold for turkey bedding. The second year he saw many kinds of brushy plants and larger blueberry bushes full of bunches of wild blueberries. By the third year, the resilient jack pine seedlings were sprouting up by the thousands.

Thirty-three years later, you would not know a fire had occurred there. Nothing is left of the fire's devastation except for a blackened stump here and there. The dead oaks have regenerated from their roots into new trees, destroyed red pine plantations were cut and replanted, and the fire-loving jack pines that burned have regenerated because of their tough cones that opened and scattered seeds by the thousands. In 2000, the Fire Hill Golf Course was carved out of the wooded acreage on Hill Road near what was the north end of the blaze.

The Five Mile Tower Fire took many miles of fire lines and many thousands of gallons of water to extinguish. However, what came after could only be described as a watershed—or perhaps waterfall—of reviews and changes. Rangers from throughout the state reported that the Five Mile Tower Fire was the agent of change. It sparked extensive sessions of critical analysis and soul-searching. Experts interviewed for this book agree that the fire led to hundreds of changes that can be seen to this day in firefighting philosophy, management, communications, and equipment.

In many human endeavors, sometimes it takes a tragedy or a series of unpleasant events to usher in a new beginning, a better system, or an updated organization with modernized goals and objectives. There were 11,744 forest fires fought by the DNR in combination with rural volunteer fire departments in Wisconsin in 1976 and 1977—an average of 16 fires each day. These fires burned a total of 119,288 acres—186 square miles of Wisconsin woodlands reduced to blackened stumps. These two years of losses made everyone from the governor on down take notice. In every region of the state, forestry leaders sat down during the late summer of 1977, studied the fires of the previous two years, and listed their recommendations for improvement.

While changes in firefighting obviously were needed, management of the Five Mile Tower Fire itself was handled so well that it became an example of excellence at the National Advanced Resource Training Center (NARTC) in Arizona. For several years, it was used as a training exercise in the Advanced Forest Fire Behavior Officer course.

Wisconsin's training and ability surprised fire managers all over the country, especially in the West. The rate of spread of the Five Mile Tower Fire of one mile every thirty-five to forty minutes was faster than even the massive fires in the western wilderness. Fire experts from California, Alaska, Idaho, and other states studied the Wisconsin blaze. They determined that, given the fuel types and weather conditions, this fire could have been many times larger. They were impressed that it was contained in such a relatively small area. They also found that the firefighting effort was well organized and the cost of suppression was just a fraction of their estimates.

Still, there was room for improvement, as the next section of this book underscores.

Part 2

Fighting Forest Fires in the Twenty-first Century

Drought, forest stand age, and other factors easily made 1976 and 1977 Wisconsin's most fire-prone years of the second half of the twentieth century. In those two years, the state had more than 11,744 wildland fires. They were widely distributed across the state's forested lands, and some of them were very large. The happy news is that Wisconsin had no forest-fire fatalities in 1976 or 1977. But damage to property and resources was significant.

The Five Mile Tower Fire had validated the DNR's work in training not only its own personnel but also local volunteer fire departments to battle a common enemy. But this devastating fire also showed that both DNR and volunteer departments had a long way to go in controlling forest blazes. Many town residents realized their fire departments needed better equipment, more protective clothing, and better radios. Some towns that didn't have fire departments decided they needed them; two of those involved in the Five Mile Tower Fire, Chicog in Washburn County and Wascott in Douglas County, immediately started the process of creating a volunteer department.

This intensive review process produced, in September 1977, a twelve-page DNR publication, *Fire Review Findings*. In it the DNR recommended a total of thirty-eight major changes, including these:

- Better facilities for headquarters and more mobile phones in vehicles
- More shift changes to reduce fatigue
- Uniform fire maps for DNR and volunteer fire departments
- Posting and record keeping of weight capacity of bridges
- Portable light kits for dozers
- More training for overhead staff for DNR employees outside of fire management
- Additional training for fire departments
- Training for new equipment
- Improvements to township and county road maps
- Radio communication between the DNR and local fire departments
- More portable radios
- Better and more powerful dozers and trucks

On October 7, 1977, DNR secretary C. D. Besadny approved all thirty-eight recommendations in *Fire Review Findings,* ushering in a new era in Wisconsin wildland firefighting.

In addition, the Wisconsin legislature enacted new laws to help protect forestlands. In 1977 railroad fire laws were updated to eliminate most references to steam locomotives and logging railroads. And the law better defined how railroads were to manage fire prevention, improving their methods of reducing sparks cast from the trains' steel wheels and burning chunks of carbon belched from smokestacks. The legislature also changed the requirements on clearing rights-of-way and gave forest rangers and other enforcement personnel more authority to stop trains and deal with forest fires started by trains. The section of the statutes on civil liability for forest fires was also updated and strengthened. State laws, improved safety handbooks, and new training have all helped reduce the number of forest fires caused by railroads.

Significant public expenditures resulted as well. A total of $1.5 million of state money and $311,700 in federal funds were spent on the following DNR forestry equipment purchases delivered to the LeMay Forestry Center in Tomahawk in January 1978:

- Fourteen tandem-axle trucks with 1,000-gallon water tanks
- Seventeen tandem-axle trucks with 850-gallon water tanks

- Twenty-six John Deere 450 bulldozers with plows and 130-gallon water tanks
- One Bombardier Muskeg low-ground-pressure unit with 200-gallon water tanks
- Nineteen three-quarter-ton pickup trucks with 150-gallon water tanks
- Thirty-one twelve-ton tilt-bed trailers to haul the JD 450 dozer-plow units
- One sixteen-ton tilt-bed trailer to haul a government excess low-ground-pressure unit

Policy changes came, too. Madison DNR headquarters officials analyzed the fire-prevention efforts of the past, and provisions were put in place to enhance forest-protection law enforcement and encourage more frequent suspension of burning permits. Fire bans were to be used more often in extremely dry periods. In times of extreme fire danger, red flag warnings are imposed outlawing any smoking, chainsawing, barbecuing, or campfire building in the area.

Since 1977, Wisconsin has not seen a fire as large as the Five Mile Tower Fire. But if the conditions are right and a fire once again gets out of control, the DNR, local volunteer fire departments, and government officials believe they are ready. From the ashes of the monster fire at Minong rose a twenty-first-century approach to fighting large-scale blazes.

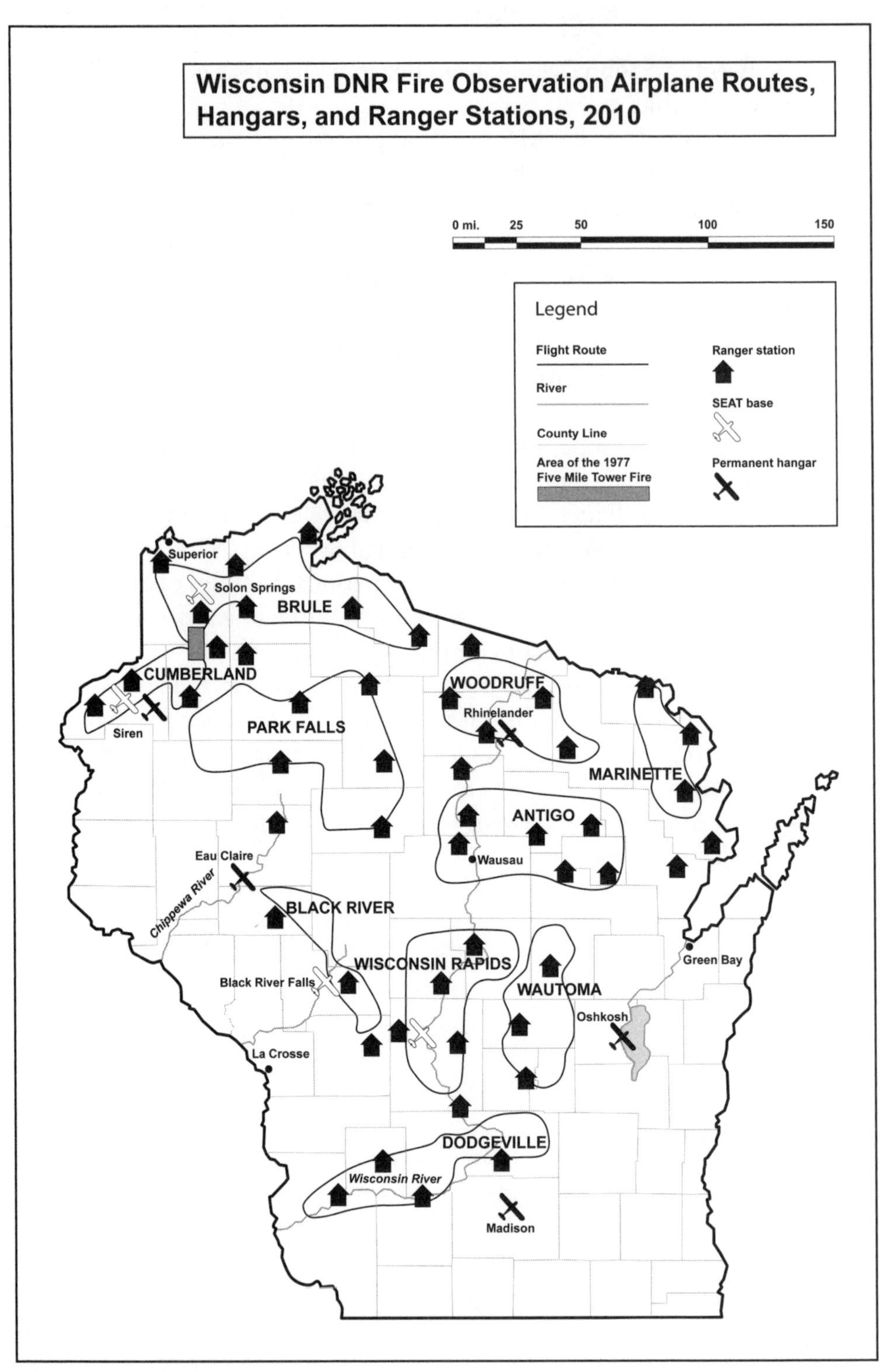

Map by Joel Heiman

Today's Ranger Stations

Wisconsin DNR ranger stations have changed greatly since the fire years of the 1970s. Today's stations are the nerve centers of each of the state's fifty-six fire-response groups. The fifty ranger stations contain high-speed computers with sophisticated software programs in communications; weather data collection; mapping, including both Geographic Information Systems and aerial photo capabilities; equipment inventories; and training. Hand-held and truck-mounted Global Positioning Systems provide accurate and reliable information to teams in the field.

To assess the threat of fire, each ranger can check weather and burning conditions online. DNR leadership has set up an instantaneous weather system that compiles data from twenty-three automated weather stations in all parts of Wisconsin, providing the rangers with more than seventeen types of data to help them plan for the day's activities. Fire-danger warnings, burning bans, and red flag warnings begin with information from this system. Complicated fuel moisture readings allow rangers to assess the rate at which a fire will spread on a particular day given other factors such as wind speed and direction.

Almost all of the ranger stations have a public counter where staff answer questions, sell licenses, give advice, issue burning permits, meet with forestland owners, and hand out literature. Most have displays of photos of forest fires as reminders of the size and scope of the challenge.

The typical DNR station contains the offices of one or two college-degreed forest rangers. Some have a forestry team leader. One or two

The Gordon Ranger Station is amply equipped with firefighting equipment, including two large Type 4 engines pulling tilt-belt trailers with John Deere 450 tractor plows and three Type 7 pickup trucks with slip-on tanks and hose reels. From left to right: John Glindining, forest technician; Rod Fouks, Douglas County team leader; Dan Jones, forestry technician; Mark Braasch, Gordon forester-ranger; and Don Luebbe, DNR forester. *Photo by Bill Matthias*

conservation wardens have their offices in the station but are usually out in the field checking on anglers, hunters, and trappers. Other personnel on duty are equipment operators and maintenance workers, who not only drive the trucks and dozers but know how to fix and maintain complicated equipment so everything is ready to go at a moment's notice. During the busy fire seasons, some stations have a part-time office manager to answer phones and help the public.

Fire-retardant uniforms are hanging at the ready, along with two-way radio sets, helmets, gloves, and maps. On fire-danger days, many rangers wear their fire clothing, except helmets and gloves. They keep their radios strapped to their chests, some with headsets or shoulder clips for microphones that can be used hands-free in the field.

Communication, including the radio transmission system, has improved vastly since the Five Mile Tower Fire. Radios have more than twenty frequencies for communicating with other DNR firefighters, equipment operators, headquarters, incident command centers, wardens, police, and dispatch centers. Radio towers are scattered across the state to "repeat" and strengthen radio signals and make communications nearly seamless. Radios can be quickly reprogrammed to switch from one tower to another. On a single fire, many radio frequencies are utilized at the same time, depending on what job is being done by the individual. Some channels have a limited range and do not use repeater towers; these are reserved for rangers, equipment operators, and fire departments on the ground fighting the blaze within a few miles of each other. Other frequencies are reserved for the leadership team at the command center to communicate to the rangers on the fire, direct resources, or request assets from other regions, and for communications with the Madison command center. The new radio system requires training—training received by everyone from the top ranger administrators to volunteer firefighters.

Most people working on a forest fire today have the capability to communicate to almost anyone at any time. And communications are constantly being improved as radio technology advances and as funding becomes available from the DNR, Homeland Security, the Federal Emergency Management Agency, and other government entities. The systems are costly, often funded by grant writing and creative budgeting by fire officials.

When walking into the modern ranger station or volunteer fire hall, one can see the portable radios lined up like giant dominoes, all plugged into quick-release chargers ready to be grabbed in case of fire. In addition, most members of volunteer fire departments and all DNR employees in fire-control jobs carry a portable radio wherever they go. Because many of the volunteer firefighters and DNR employees also volunteer as emergency medical technicians or first responders in their spare time, these portable radios are a lifesaving tool.

Local Volunteer Fire Departments

Amazing changes have been made in fire department training, equipment, vehicles, clothing, and communications in the years since the Five Mile Tower Fire.

Today's local fire departments now work in tandem with the DNR ranger stations to put out fires. In the case of wildland fires, area fire departments are paged to the scene at the same time DNR heavy units are called. For most structure fires in rural areas where the local fire department is summoned, the DNR also arrives in case the flames spread to adjoining woodlands. It is a cooperative venture—no one goes it alone anymore. The DNR and local fire departments train together and work together. They also have common radio systems to be able to communicate with each other.

Although the DNR is still authorized to call upon "able-bodied citizens" to help in fighting forest fires, it is no longer common practice to do so. This makes the local fire departments crucial in fighting wildland fires. The fire departments not only fight structure fires, help protect structures, and assist in wildland fire suppression, but they also help evacuate citizens caught in the path of the flames.

Today's volunteer or paid fire departments have many trucks and pieces of equipment at their disposal. They have large pumpers to act as the "water servers," much as a computer server handles many stand-alone computers in an office. The pumper can take in water from lakes, hydrants, or portable ground tanks (or from its own supply) and send water out through many different-sized hoses to

The volunteers and equipment of the Wascott Volunteer Fire Department in Douglas County. Equipment seen here includes an ambulance, a wildland sprinkler system truck, an F-350 pulling fireboat with portable pump and hoses, a tanker, pumpers, an ATV with rescue trailer, and an equipment truck. EMS personnel, left to right: Betty Ebert, Brian Ebert, and Dottie Fries. Firefighters, front row: John Glindining, Stacy Vig, Steve Balcsik, Ken Mertz, and Tim Johnson. Back row: Art Lange, Tom Michalek, Bob Foster, Bill Matthias, John Almer, and Oliver "Sully" Soltau.

Photo courtesy of Steve Vogel

the flames. Another important truck is the water tender, which carries thousands of gallons of water and can dump the load of water into a folding ground tank and then shuttle to a "water point" to be quickly refilled and returned to the fire to dump again. This truck is valuable in rural areas where there are no water hydrants. Tenders also carry portable pumps so they can be used in some situations to squirt water, not just haul water.

Rural fire departments have established "dry hydrants" for drawing water from lakes or streams when needed to fill trucks. These do not freeze in the winter; rather, the pipes remain dry until attached to a suction hose and pump. County emergency maps show the locations of these dry hydrants, and fire departments practice using them on a regular basis. They are usually located next to a water source on a blacktop or easy-access gravel road.

Fire departments also use "fast-attack" pumpers for initial attack on a fire, for pumping water from lakes or reservoirs, or for fighting small fires with hoses. They are usually not as large as the massive pumpers or tenders, sometimes built on a three-quarter-ton or one-ton pickup truck chassis.

Many rural fire departments have a pickup truck pumper built with high ground clearances, with a small 150- to 300-gallon water tank and hoses for use in forestlands to assist the DNR in the initial attack on a forest fire.

In addition, departments have equipment for search and rescue, such as ATVs with rescue trailers, snowmobiles with rescue sleds, fire and rescue boats, hovercrafts, equipment trailers, forestland sprinkler systems on trucks, and equipment trucks hauling hundreds of pieces of firefighting or rescue equipment.

The twenty-first-century firefighter has two complete sets of clothing for fires: one set for structure fires that can withstand fourteen hundred degrees of heat and one lightweight set of Nomex for wildland fires. They also have two helmets and two pairs of boots.

Engines, Dozers, and More

State laws provide forest rangers with considerable authority during a fire, not unlike military generals in times of war. Rangers have the authority to request or recommend evacuations, close roads, encircle the zone with police, and fight the fire with all assets necessary. The ranger in charge, or incident commander (IC), has the power under state law and in accordance with the accepted protocol of fire training to be the "CEO of the fire scene." All volunteer fire departments called, all other DNR personnel dispatched, all law-enforcement officers, and all private contractors with equipment must follow the dictates of the DNR fire-control officials.

Wisconsin's policy calls for a rapid initial attack with as much equipment and personnel as are needed to control the fire. In the event of a project fire, like the Five Mile Tower Fire, the policy also says "we will fight the fires we have" and not worry about the possibility of another fire. All DNR forest-fire-control personnel are focused on this one goal.

Trained people are the most important element, but quality equipment is crucial, too.

As firefighting equipment goes, the Type 4 engine is the workhorse of the DNR ranger station. It is a large truck painted lime green with an inventory list of more than seventy-five pieces of equipment.

The Type 4 is much like a volunteer fire department primary pumper, except it is also used to tow a Type 5 tractor/dozer/plow on a tilt-bed trailer. When this engine comes down the road pulling a trailer with a John Deere 450 tractor/dozer/plow, it's a formidable sight. The truck also carries 850 gallons of water weighing three and a half tons, several pumps, and hundreds of feet of fire hoses. It can apply water on wildland fires or structures and can draft water out of streams or lakes for its own uses or for supplying to other trucks or filling backpack cans for hand crews.

The Type 4 truck has foam equipment to add chemicals to the water to "foam" fires, which stretches the water supply by cutting down on evaporation. The foam also blocks oxygen to the fires, helping firefighters more quickly knock down flames.

This highly adapted truck includes many compartments containing everything from flashlights and chain saws to portable pumps, hand tools, spools of hose, and hundreds of tools and connectors needed for firefighting.

The John Deere 450 is *the* tractor/dozer/plow for fighting forest fires. Wisconsin can field about eighty of these JD 450s at any given time. A six-foot blade on the front of the unit can push over trees as big as nine inches in diameter, and a plow on the rear enables it to plow a wide furrow at the same time. The rear-mounted plow digs a flat-bottomed ditch six feet wide down to mineral soil.

These huge machines cut firebreaks, also called furrows. If two or three JD 450s get on a fire quickly, they can work side by side to make a firebreak more than twenty feet wide. If the fire is running too fast, the units will stay on the edges to pinch the blaze to a point somewhere in the distance.

The JD 450 has hydrostatic drive with an eighty-five-horsepower engine. It hauls 150 gallons of water and two hundred feet of one-inch hose for fighting small fires or protecting the operator in the flames. In older models, a water shower curtain can be activated in an emergency, causing a cascade of water to spray down on all sides of the operator's seat. The driver also can pull down quilted thermo-protected curtains for added protection in hot places.

The cabs on newer JD 450 are tempered glass enclosures with air-conditioning and heat; they do not need the shower system or the quilted blanket protection. The tempered glass has been tried in fires and passed the test of one-thousand-degree heat, allowing the operator to keep plowing lines while safe inside the enclosure.

Two John Deere 450 tractor plows work on the Cottonville Fire in Adams County, Wisconsin, May 5, 2005.

Wisconsin DNR file photo by Mike Lehman

The tractor-plow unit has a thirty-five-gallon fuel tank, allowing it to work on fire lines for more than ten hours without refueling. The unit also contains drinking water for the operator.

During the Five Mile Tower Fire, some dozer operators did not have radio contact with fire officials. They had to get their orders personally from the rangers shouting over the engine and fire noise. The modern operator has a safety helmet with a radio system built in, much like an airplane pilot's. This allows the driver to communicate while using both hands in the operation of his JD 450.

In 1977 some dozers did not have protective cabs over the operator or lights for nighttime use. Now all DNR JD 450s provide full protection for the driver and lights for fighting fires all night. In addition, the DNR has portable light kits that can be clamped on private dozers called to a fire.

When several JD 450s are used in tandem with larger D-6 or D-7 dozers, drivable roads can be quickly constructed on the edges of the flames. These drivable roads through the dirt and sand allow four-wheel-drive pickup trucks to come behind to spray the flames or patrol fire lines. The JD 450 is sent out first, making the furrows that stop the flames. Then the dozers come behind, cleaning up the downed trees, pushing brush to the sides, and taking out stumps and roots.

The ranger's initial attack vehicle is a Type 6 pickup truck or engine, an indispensable tool. Each ranger, team leader, and regional supervisor has a three-quarter-ton pickup built to heavy-duty specifications and painted lime green. The truck has a six-foot box with a water tank and pump assembly system along with a reel of hard-rubber fire hose. In addition, there is room for carrying several backpack spray cans or bladders, extra-small portable pumps, and hundreds of feet of cloth fire hose to stretch long distances as needed to reach flames. Many extra compartments are built into the truck body to haul equipment, fittings, connectors, chain saws, and extra water. An emergency light system and siren horns are atop each truck.

Inside, the ranger has map kits, plat books, emergency government county highway and residential maps, forestry maps, and special DNR detailed maps showing fire lanes, hiking trails, four-wheeler and snowmobile trails, and private landowner access roads.

A Modern-Day Fire-Management Team

It takes a community of dedicated individuals to put out a fire. People with a passion to protect natural resources from top to bottom serve Wisconsin's citizens in a portion of the DNR called the Bureau of Forest Protection. This group includes a director and three section chiefs: wildland fire/equipment research and development, fire management, and aviation.

The DNR's fire-management section is located in Madison but affects every ranger station in the state. The section manager oversees the state's fire program and connects with fifty-six fire-response rangers, five regional forestry leaders, nine area forestry leaders, and specialists in the Madison, Woodruff, Poynette, Montello, and Wisconsin Dells offices.

The section manager is leader of the Wisconsin Interagency Fire Council, comprising other fire administrators from the fish and wildlife division of the DNR, the United States Department of Agriculture Forest Service, the Department of the Interior Bureau of Indian Affairs, and the National Park Service, along with Menominee Tribal Enterprises. They work together to provide cross training and strategic fire planning.

The section manager also represents the state on the executive committee of the Great Lakes Forest Fire Compact. This group, which includes Minnesota, Michigan, and Wisconsin as well as the provinces of Manitoba and Ontario, agrees to cooperate in large fires and provides training opportunities, shares ideas and personnel, and exchanges equipment. This is the group from which Wisconsin obtains firefighting aircraft if needed for large fires.

For firefighting purposes, the state is divided geographically into five regions, each with one forestry leader. As an example, the northern regional forestry leader in Spooner is in charge of thirty-eight supervisors, specialists, and rangers. This northern region is divided into four areas, each with an area forestry leader in charge of team leaders. Each team leader in turn supervises from three to five ranger stations. In each ranger station are a ranger and several technicians. Generally, the regional leaders, area leaders, and team leaders will be located in the incident command centers during a fire, working in a leadership capacity.

The specialists provide specific services to the wildland firefighting community and the public. They include a suppression specialist, a cooperative fire specialist, a wildland urban interface specialist, an operations specialist, two forestry law-enforcement specialists, a wildfire-prevention specialist, and a part-time railroad fire-prevention and locomotive-inspection specialist.

The Division of Forestry training section administers the annual training of all rangers and forestry technician/equipment operators. Training is a major component in the mission of fire management, keeping everyone up to date in fire management, organization, laws, and practice. In addition, the fire-control personnel at the local level provide training to volunteer fire departments, other cooperators, and the emergency fire warden force.

In the end, the whole system relies on training. The philosophy is to train everyone involved on a fire until each person knows his or her job—and then train some more. The system relies on cross-training, so that personnel are proficient in several different positions. Training for safety and effectiveness is continual and will always be the top priority for DNR rangers, fire chiefs, emergency government officials, firefighters, and equipment operators.

LeMay Center in Tomahawk

Hidden off busy Highway 51 in Tomahawk on an ordinary-looking city street called Somo Avenue is a group of brick and metal buildings, some dating back to the early part of the twentieth century.

This complex is the DNR's LeMay Forestry Center, named in honor of Wisconsin's chief forest ranger Neil LeMay, who served from 1942 to 1969. The outside may look a bit timeworn, but inside is state-of-the-art technology.

Wisconsin's fifty-six DNR fire-response ranger stations house a fleet of Type 6 and Type 4 engines plus trailers, Type 5 tractor/dozer/plow units, and specialized equipment such as low-ground pressure tractors for use in marsh and bog areas. The Tomahawk center

At the LeMay Forestry Center in Tomahawk, Section Chief Mike Lehman stands with three John Deere 450 dozers that will be fabricated for firefighting.

Photo by Bill Matthias

services them all. Every piece of equipment—chassis, truck, plow, dozer, pump, hose, fitting, coupling, to name a few—passes through the center, where it is designed, fabricated, fitted, welded, bolted, and retrofitted as necessary, and then moved out. The total equipment inventory numbers in the thousands. There is also a small cache of engines and dozers kept at Tomahawk to provide replacements for broken or damaged units or as reinforcement on project fires such as the Five Mile Tower Fire.

A staff of fourteen includes the section chief in charge of wildland fire equipment; nine expert technicians with skills in welding, plumbing, electrical and mechanical fabrication, and repair; storekeeper; purchasing agent; mechanical designer; and office secretary. With a budget and inventory in the millions of dollars, the dedicated personnel in this old building prepare and help maintain the equipment for forest protection for the entire state.

The center uses advanced computerized design and drafting technology complemented by computer-controlled digital numerical milling machines and lathes. Personnel use computer-guided laser-plasma cutting machines to make signage for the equipment and to make large signs used throughout the state by rangers and fire wardens.

A large stockroom contains parts, equipment, and clothing necessary to supply the fifty-six DNR ranger stations. Lined up on heavy shelves and on the floor are subassemblies of water pumps, hydraulic fluid pumps, hoses, portable light kits for private dozers without lighting, spare clothing, extra dozer doors, water manifolds, and several slide-in tank-and-pump assemblies with hose spools ready to be attached in the bed of three-quarter-ton pickups. Large trucks and trailers are ready at a minute's notice to move out and resupply a large fire and replace and repair broken equipment day or night. One cache trailer has everything necessary for two twenty-person firefighting crews, including fire-retardant clothing, portable pumps, hoses, and other equipment needed along the fire line. The trailer also holds portable work lights, an air compressor, hand tools, and two four-stoke gasoline-powered portable pump kits. An equipment and repair truck is also at the ready, containing generators for nighttime lights, mechanics' tools, welder, crane, spare parts for the dozers, pumps, hydraulic equipment, and a spare fire plow for the JD 450 dozer, as well as a spare plow depth adjuster. This truck and a crew of two mechanics can go out to a forest-fire scene and do repairs in any location, day or night.

In the fabrication building, several John Deere 450 dozers can be outfitted simultaneously with attachments necessary for firefighting. At times it looks like a manufacturing plant, with the dozers lined up in various stages of completion and prefabricated hydraulic pumps, hoses, water pumps, water tanks, and chain boxes organized on the floor and ready for attaching to the dozers. These dozers are replaced about every twenty years, so each year five or six of the pool of dozers need to be purchased and outfitted.

The LeMay Center also outfits the various types of engines with toolboxes, radios, red lights, winches on some units, water tanks, foam systems, pumps, and hoses.

For wetland fires, the center has low-ground-pressure tracked vehicles outfitted with water tanks, pumps, foam supplies, and hoses. These are available for firefighting in the hard-to-reach swamps, bogs, and wetlands where dozers would get stuck or bogged down. During the Five Mile Tower Fire, several dozers got stuck and had to be pulled out. In other cases, fires in swampy areas have had to be fought with hand labor. These units provide new flexibility for firefighting in marsh, swamp, and other wetland areas.

Research and development is ongoing at the LeMay facility. Staff members design and test parts and components for each firefighting apparatus with efficiency and safety in mind. They test equipment in the field in live conditions and share knowledge with other states and provinces regarding new ways of doing things or utilization of new pieces of equipment.

From 1934, when the DNR equipment center was established, to the present, this research and development arm of the DNR Forestry Division has served the state of Wisconsin well. Its dedicated personnel have made the forestlands of the state a safer place to work and play by continually upgrading and improving the firefighting equipment.

Madison Command Center

In some fires or natural disasters, such as floods or tornadoes, the DNR director of the Bureau of Forest Protection and the chief of forest-fire management may determine it is necessary to activate the Madison Command Center, which serves forestry, law enforcement, parks, and water programs by coordinating the movement of personnel and equipment from across the state to meet the needs of incident commanders in the field. Officials there can also order personnel or equipment from other states or programs to assist and, if necessary, assign priorities for the dispatch of personnel and equipment.

The current Madison Command Center is a large room in the lower level of the current DNR headquarters, two blocks off the Capitol Square in Madison. Command center staff can instantly hook up laptop computers, phone systems, and radio systems to be in contact with the leadership team on the scene of a fire or other disaster. Four desks are bristling with phone lines, cables, and computers, and the walls are lined with eight-foot maps of each dispatch center of the state. Magnetized indicators can be moved about, tracking personnel and equipment. Whatever the incident commander asks for, the Madison office can deliver.

The bureau chief of forest protection and the section chief of fire management can visualize assets or holes in the protection umbrella and marshal resources accordingly.

If the rangers and heavy units in an area are already on-site and more resources are needed, the Madison Command Center leaders begin calling the regional forestry leaders around the state to move personnel, equipment, trucks, dozers, and plows to fill in behind

At the Madison Command Center, Trent Marty (director, Bureau of Forest Protection, Division of Forestry), Mary Roberts (DNR assistant and daughter of File Mile Tower Fire line boss Tom Roberts), author Bill Matthias, and Blair Anderson (chief, forest-fire management, Division of Forestry) examine the DNR maps of the Five Mile Tower Fire.
Wisconsin DNR file photo by James R. Miller

units that may already be on the fire or en route to the incident. This "move-up" or "fill-in" method provides for an orderly transition and the flexibility to divert units to other incidents if needed. It is also in keeping with the DNR's "we will fight the fires we have" philosophy. The command center can also call upon the governor if necessary for deployment of National Guard equipment, ranging from fuel haulers to helicopters, to assist in the effort. Under the Great Lakes Forest Fire Compact, officials can also request the help of CL215 and CL415 water-dropping bombers or other equipment or personnel from Minnesota, Michigan, Ontario, or Manitoba.

The command center allocates resources anywhere in the state, assists in the initial attack, contacts the five regional forestry leaders, and helps mobilize resources. These five regional forestry leaders then contact the nine area forestry leaders. The LeMay Forestry Center in Tomahawk is contacted for resources ready to be moved to any large fire. The Eastern Area Coordination Fire Center in St. Paul, Minnesota, is contacted for information on availability of Minnesota's airplanes. Four firefighting planes are available from Minnesota and ten from Ontario and Manitoba.

Instant weather data is accessed at the command center from the twenty-three DNR weather stations around the state. They provide live updates on wind speeds, burning index numbers, ignition components, and fire-spread mathematical predictions.

Possibly one of the most important roles the Madison Command Center plays is to provide for personnel safety. By moving people and machines toward the fire from distant locations of low fire danger, extra rangers and equipment operators will be available to give the initial attack people rest after eight or ten hours. That's a vastly different scenario from on the Five Mile Tower Fire, when a lack of rangers and operators required keeping personnel on duty for twenty-four, forty-eight, even sixty or more hours without sleep.

What DNR leadership attempts to do today is provide enough personnel so crews gets reasonable rest, meals, and liquids, as well as sleep between shifts. An important component of adding personnel is the modern addition of trained hand crews throughout the state available to be sent on large fires. These fire crews are available from the University of Wisconsin–Madison, UW–River Falls, UW–Stevens Point, Northland College in Ashland, the DNR's south central region, Gordon Prison Camp, Black River Falls Prison Camp, Fox Valley Technical College in Appleton, and a Mennonite crew from Park Falls. Some rangers are organizing hand crews of trained eighteen-year-old high school students in pine regions of the state.

The command center can access two twenty-person sets of fire clothing and gear from each of the nine areas of the state to outfit local firefighters or local officials in the fire region. All of the volunteer fire departments in Wisconsin's fire-prone regions receive DNR wildland firefighting training and possess the proper Nomex fire-retardant clothing, hard hats, gloves, leather boots, emergency gear, fire shelter packs, water supplies, and portable radios. Command center officials can also call volunteer fire departments from long distances away to give relief to the initial attack departments after a few hours of heavy labor in hot, dangerous conditions. These departments can be on the road traveling toward the fire during the first hours of the fire to provide support later if needed.

The Madison center is staffed by four or five individuals around the clock during major fires or incidents. Staffing includes the section chief of fire management, the bureau chief of forestry, a plans chief from the fire-management staff, a resource unit leader from the forestry division, and a public information officer in charge of communications and the media. Other employees in the Madison offices of the DNR may also be asked to help, and there is a special seat for the governor if he or she wants to stop in for updates.

Within a few years, advanced touch-screen, microwave radio communications will link the Madison Command Center with every one of the DNR dispatch centers and ranger stations.

Incident Command Post

The Five Mile Tower Fire was fought from a farm field and managed from the front seats of pickup trucks for three days. Now, large fires are fought from prepositioned, preplanned, and prepackaged command centers. Within the first hour after a fire is designated a project fire, an incident command post (ICP) is chosen near the origin of the fire. There the incident commander—once called the fire boss—works in concert with the other leadership personnel.

DNR rangers, along with emergency government officials and fire chiefs, have selected various locations around the state—such as local fire halls, schools, and municipal buildings—for future incident command centers. These buildings have been outfitted so that they can be quickly converted into ICPs. Sites were selected because of their locations near pine forests and because they have good road

Wisconsin DNR and local fire department staff work in the incident command post set up for a mock project fire in Barnes, Wisconsin, 2008.

Photo by Cathy Khalar, Wisconsin DNR

access and adequate parking. These buildings have bathrooms and drinking water, and most have kitchen facilities, storage rooms, and space for meetings or gatherings. They have communications lines and other necessities.

These facilities have been "beefed up" electronically and technologically to provide everything an overhead team needs to lead the fight in a project fire. A tall antenna was built next to each building, with a large cable junction box on an inside wall for attaching multiple radio devices. Most locations have six extra phone lines that can be activated within minutes by the phone company. The posts also have six antenna leads to which radio sets can be attached.

The overhead leadership team requires many positions in a large fire situation. Tables are set up with plenty of chairs. Radio sets, cables, wires, and laptop computers instantly make the place look like a school classroom for communications technology. Within sixty minutes a once-quiet fire hall becomes the nerve center of a major forest fire, with vehicles clogging the property.

The incident commander is in charge of the entire operation. Four major centers are organized within the building space. The plans division is in charge of checking in personnel and equipment and of all of the resources. The logistics division calls for added support. The operations division has four directors: structural branch director in charge of the fire departments trying to save structures; wildland branch director, who assesses forest types and helps plan where to send assets in the forestlands; law-enforcement branch director, in charge of wardens and police units for road blockage, evacuations, and mitigating looting; and air operations director, in charge of spotter planes and water bombers. The fourth center is command.

All of these management units are trained to work in radio contact with the rangers and fire personnel in the smoke and heat. With accurate maps and knowledge from spotter planes, the incident commander and other fire directors can follow the fire foot by foot.

During the first hour of the fire, a protocol that has been rehearsed and honed with many hours of simulated training begins. The first ranger on the fire will start as the incident commander, manage the first attacks, and call for additional manpower, DNR equipment, and fire departments. The next two rangers on the fire scene become the "line construction" or division supervisors, one on each side of the

Volunteer fire trucks from many departments stand ready to be deployed to a mock project fire, Barnes Ranger Station, 2008.

Photo by Cathy Khalar, Wisconsin DNR

fire. They supervise the heavy units and firefighters, starting from the origin of the fire and constructing fire lines on both edges of the flames.

If the fire continues to flare ahead, the single-engine air tanker (SEAT) plane is dispatched as the initial-attack water-dumping plane.

The fourth ranger on the fire will most likely be the area supervising ranger, who will take over the duties of incident commander, allowing the first ranger to be the H Division boss. He has the most knowledge of the area and will constantly use firefighters and equipment to try to stop the forward-moving flames. The H Division boss will always keep heavy units and dozers on their trailers ready to unload when a situation occurs that may allow the fire to be fought at the head.

At this time, with the fire not yet under control, the area ranger will pick a nearby incident command post and establish a command center. He will leave the fire scene, drive to the ICP, and call for additional personnel to arrive and begin the fire organization.

How does the incident command post get set up so fast? Usually, the initial attack ranger triggers setup when he calls for the activation of an overhead team with the necessary personnel and equipment to fight the fire. The incident command equipment trailer and van pulling it will get on the road and within an hour will arrive at the command center with everything necessary to turn the building into an instant communications hub. Usually, part-time DNR employees working during high-fire-danger weeks will be on duty to drive this equipment trailer to the designated site.

There is a three-ring binder for each predetermined center with drawings of the facility and sketches showing where everything should be placed. In the trailers are maps, drawings, directions, copy machines, telephones, laptop computers, cords, cables, printers, extra portable pumps, hoses, sprinkler kits, water, ready-to-eat meals, radios, portable radio antenna masts, signs, tables, and chairs. The directions and sketches make it possible to set up the command center in a matter of minutes.

Each area of the state has a fire department mobile-command van loaded with communications gear for use by the local fire department boss in and around the fire scene. The first fire department chief on the fire is usually assigned the duties of the fire boss and gets into this mobile-communications van along with a driver. This van goes in and around the flames, organizing the many fire trucks arriving on scene to save structures. The van has a table and chairs with maps, radios, and antennae for both low-band and VHF communications. This fire boss is the eyes in the field for the structural branch director back at the incident command post, directing and advising asset placement as well as stressing safety for firefighters. Those in the van will scout for places to send trucks and personnel in order to attempt to save structures without endangering the lives of firefighters. The vans also contain extra wildland fire clothing for added safety of firefighters.

The command van is used by many fire chiefs in rehearsals during mock fires. These practice sessions are invaluable for efficient use of equipment and better firefighting when the real thing occurs.

Some areas of the state have built a large trailer or truck into an incident command management team center. The Douglas County Office of Emergency Management, for example, has built a complete command center that can be used anywhere, even in remote locations where there is no electrical power or telephone service. From this trailer emergency management officials can help implement evacuations, facilitate sheltering of displaced or burned-out home owners, activate health and human services assistance to firefighters and citizens, and request Red Cross and Salvation Army assistance. They have computerized phone systems that can activate recorded emergency messages to people in harm's way. These messages can be sent by township, legal description, section numbers, or radius distance from a given location. It is another safety net for the citizens of Wisconsin to receive notices by phone about impending danger.

The command vehicle is a big truck, thirty feet long. It is loaded with a twenty-five-kilowatt generator powerful enough to support not only the command truck/trailer but also an incident command post building in case of power outage. Other supplies on board are radios, cameras, floodlights, office equipment and desks, computers, telephones, and a portable radio tower that can be snapped into place.

Even if it is not used as the incident command post, the truck can be parked next to the building to provide extra emergency assistance with added phones, radios, electrical power, and outside lighting. It is an impressive, extremely valuable tool.

DNR Air Section

In forested regions of the state, the DNR air section is alert and watching. Officially called the DNR Aeronautics Section, it is part of the Division of Forestry, Bureau of Forest Protection. One part of its mission is firefighting and helping efforts to put out forest fires.

Whenever it is dry, especially in the spring of the year when the woods have not greened up and the fire danger is moderate to high, look up in the sky. It's likely you will see a high-winged Cessna airplane flying a quarter mile above the ground. The pilot and observer are up there patrolling the skies of Wisconsin looking for "smokes."

The single-engine airplanes flying high in the air, working with the operational fire towers, are the front line in the effort to fight forest fires. If citizens don't report a fire, chances are it will be seen in minutes by some DNR employee, either swaying side to side in the

A DNR spotter plane sits ready for takeoff at the Solon Springs airport.

Wisconsin DNR file photo by Mark Braasch

top perch of a tower or bumping around in the hot air turbulence in the tight cockpit of a small airplane.

Take a Wisconsin map and draw ten circles that together cover the state, and you have a map that looks like the one on the wall in the Madison DNR hangar of the aeronautics section chief. The state has been divided into ten fire-patrol routes. One aircraft is assigned to each route, and ten full-time pilots fly these routes whenever there is the potential for forest fires. During high-fire-danger weeks, ten additional part-time pilots are available. These planes are located in hangars at key locations around the state for quick access to any location in Wisconsin.

The Wisconsin state hangar in Madison supports one pilot and one fire route. Oshkosh hosts three pilots and three routes; Rhinelander, two pilots and two routes; Siren, three pilots and three routes; and Eau Claire, one pilot and one route.

These experienced pilots must be able to fly in dangerous conditions in almost any weather, sometimes low to the ground and in smoke. They are up there bouncing around in the hot updrafts in the four-seat airplanes looking down, looking sideways, following their instruments on the dashboard, and all the while communicating with the area dispatchers, rangers, or incident commanders on the ground.

The DNR pilots must also act as air-traffic controllers when other firefighting aircraft are called in on large fires. Those planes come from all sides, and DNR observation pilots have the tough task of managing the skies. The DNR patrol planes serve as the air-controlling platform in case a fire demands water dumping by special aircraft called in for big fires. State pilots rise high in the sky to stay separated from the water bombers and to be the eyes and ears of the incident commanders on the ground as well as the other pilots. The colored slurry needs to go where needed but the airplanes must remain separated at safe distances from one another and from routes to and from the water sources. Safety for the pilots and ground firefighters is of paramount importance.

Each year, the DNR pilots receive weeks of training to review correct procedures, go over new equipment, and practice with updated radios and GPS systems. They experience different scenarios and simulations.

As a member of the Great Lakes Forest Fire Compact with other states and provinces, Wisconsin is the specialist in heavy-water-tanker engine trucks and the John Deere 450 dozers. Other states and the provinces have extensive water-dropping airplane resources, and they share these resources as needed.

New chapters in fighting Wisconsin forest fires include specialized aircraft not available during the Five Mile Tower Fire. The SEAT, the CL215, and the CL415 are powerful new tools, dropping water from the sky to assist the firefighters on the ground.

The SEAT is the "single-engine air tanker" made expressly for bulk water hauling and dumping. These brightly colored airplanes look as though they are all engine and wings. But the fuselage hides an eight-hundred-gallon tank in its belly that can carry sixty-four hundred pounds of water mixed with thermo-gel and a blue dye. The nose sits high in the air with a large, five-bladed propeller and a turbocharged engine with fifteen hundred horsepower. The engine is manufactured by Pratt & Whitney, and the plane takes off from the ground at 90 miles per hour. It dives and drops its load only sixty feet above the ground at 120 miles per hour.

The pilot can not do his job without the ground-support manager. When it's time to refill the plane, the pilot stays in the cockpit while

Pilot Dutch Snyder (left) and ground support Jim Lincomfelt with a SEAT turboprop at the Solon Springs airport.

Photo by Bill Matthias

A SEAT plane drops blue slurry on a Douglas County fire.

Wisconsin DNR file photo by Rod Fouks

the ground-support manager fills the plane with eight hundred gallons of water from two large "pumpkin tanks" (which look like giant, sawed-off pumpkins), along with ten gallons of thermo-gel, plus blue dye. A large flatbed trailer parked nearby holds jet fuel for the two wing tanks on the plane, a slurry pump, a water tank, a generator, a power sprayer, and three-inch hoses for filling the plane.

The thermo-gel acts as a chemical agent allowing water to "sit" on the ground like foam, which puts out flames better than water alone. It has been proven harmless to the environment, animals, and humans. The blue dye coloring makes it visible from the air or ground so firefighters and pilots can see where the load was dumped.

Each April and May, the DNR contracts for the use of several of these SEAT planes from a private contractor. They are stationed at airports in key locations near the highly fire-prone sandy soil and pine regions of the state. The pilot and ground crew usually are on duty waiting for forest fires for twelve weeks during high-fire-danger season.

The CL215 is a two-engine water bomber, larger than the SEAT plane, designed for fighting forest fires in remote regions of Canada but now also used in the United States and other parts of the world. The CL415 is also a two-engine water-dumping plane, with newer, more powerful engines and more modern electronics and avionics. These valuable firefighting aircraft are sometimes the only means rangers have to battle blazes in certain remote regions. They are manufactured by Canadair and Bombardier, built upon request of firefighters in Quebec back in 1966. Protection doesn't come cheap: the CL415 has a hefty pricetag of $34 million. As part of the Great Lakes Forest Fire Compact, Wisconsin has access to several large air tankers in Minnesota, only an hour's flight time away from our northern pine regions.

The CL215 and CL415 are painted bright yellow with red stripes. They look like fat school buses with wings. The gutsy pilots can scoop water while skimming across lakes, rivers, or reservoirs to fill the fourteen-hundred-gallon tanks. This is a huge load for an airplane, but the two turboprop Pratt & Whitney engines, which produce twenty-one hundred horsepower each, can keep the plane flying, making it possible to fill the tanks in only ten seconds while going eighty miles per hour. The twin-engine floatplane then flies to the fire and dumps the load where requested at low altitudes and slow speeds. Piloting these craft is a very dangerous occupation.

A CL215 Minnesota water tanker scoops water to fill its tanks.

Wisconsin DNR file photo

Epilogue

In the years since 1977, there has not been another Wisconsin forest fire as large as the Five Mile Tower Fire. In monetary terms, the supression cost of the fire was $113,395, and the timber and building losses amounted to $1,177,390. In the end, the grand total loss was $1,290,785. That was a history-making year. It was one of the last "citizen fires" in which calls went out for any able-bodied person to join forces with the DNR rangers and help fight a blaze raging out of control. Men and women from ages sixteen to eighty went out in the clothing they happened to be wearing on that hot, windy day, some in T-shirts and tennis shoes, many in boots and jeans. No one had yellow Nomex fire-retardant clothing, and no one had hard hats except the DNR officials.

Today fire officials would block roads and not allow anyone into the fire scene who is not trained to fight wildland fires or who does not have fire-protective clothing and gear. The newer, more powerful and safer John Deere 450s would plow the fire lines, and the skies would be buzzing with single- and double-engine water-bomber airplanes.

But in 1977, when citizens saw smoke, they threw shovels in their pickup trucks and headed off to help. I have spent three years researching and telling this story. The editors and I have spent many hours listing the names of the people who fought the fire. Every one of the 1,610 people listed—and the hundreds not recorded—are part of history. This book is written by one person, but there are more than two thousand other stories about this fire from all sides and angles.

We simply did what we thought was right—to drive out to the headquarters, sign in, and be sent off somewhere into the flames and blackness to do hand labor. I was just one of thousands who was attempting to do what I could to bring the flaming beast under control. We did not know at the time how long it would take, how many hours we would labor without sleep, or that we all would be a part of an important chapter of Wisconsin history, fighting the largest fire from a single match in half a century.

Acknowledgments and Interview List

Thank you to the Wisconsin Historical Society for believing in my vision for this project. Special thanks to Wisconsin Historical Society Press director Kathy Borkowski and acquisitions editor Kate Thompson. Their skills and those of the expert staff of the Historical Society turned a rough draft into a finished product. I also appreciate the work of editor Bill Berry of Stevens Point, copyeditor Dawn Shoemaker, editorial assistant Mallory Kirby, and production editor Diane Drexler. Thanks also to my son Greg Matthias, who witnessed my interview with John Schultz, and to Audrey Visger, Northwoods Schools, for her help with proofreading.

Thank you to James R. Miller, railroad fire-prevention specialist, Bureau of Forest Protection and Division of Forestry with the Wisconsin DNR. Jim provided invaluable assistance, knowledge of firefighting, historical perspective, photos, documents, and information on training and weather data. Over the two years I spent researching this book, Jim Miller has been both mentor and friend.

Special thanks to four forest rangers whom I have had the pleasure of knowing for thirty-five years and who were the first persons I interviewed for this book: John DeLaMater, Ed Forrester, Bill Scott, and Barry Stanek. All four fought the Five Mile Tower Fire from beginning to end in leadership positions. In addition, I want to thank all of the current and retired DNR staff as well as citizens, volunteer firefighters, and others who were willing to be interviewed and gave of their time to tell their stories. Their names follow.

DNR Personnel and Other Fire Professionals Interviewed

Trent Marty, director, Bureau of Forest Protection, Division of Forestry, DNR Madison
Blair W. Anderson, chief, forest-fire management, Division of Forestry, DNR Madison
James R. Miller, railroad fire-prevention specialist, DNR Madison
Michael A. Lehman, chief, Wildland Fire Equipment Research and Development Section, DNR—LeMay Center, Tomahawk
James Barnier, forest-fire suppression specialist, DNR Wisconsin Dells
John A. Jorgensen, chief, Aviation Section, DNR Dane County Regional Airport, Madison
Jolene Ackerman, wildland-urban interface specialist, Division of Forestry, DNR Madison
Catherine Regan, wildfire-prevention specialist, DNR Madison
Beth Bartol, dispatcher, DNR Brule
Mark Braasch, ranger, DNR Gordon
Mike Chmielecki, fire chief, Gordon Volunteer Fire Department
Erik Finstad, shared the fire narrative written by his father, the late Roger Finstad from Gordon
Robert J. Focht, ranger, DNR Spooner
Rod Fouks, Douglas County team leader, DNR Gordon
Jay Gallagher, area forestry supervisor, Lake Superior area, Brule
John Glindining, tractor-plow operator, DNR Gordon
Larry Glodoski, area forestry supervisor, Park Falls dispatch group, Hayward
Jim Gobel, Spooner DNR Ranger Station, Spooner
Karyn Hullinger, dispatcher, DNR Cumberland, assisted me with the fire archives in Cumberland
Dan Jones, tractor-plow operator, DNR Gordon
Marty Kasinskas, LTE tractor-plow operator, DNR Gordon
Keith Kesler, Douglas County emergency management coordinator, Brule fire chief
Jim Lincomfelt, ground-support technician, SEAT plane, Solon Springs Airport
Joe Menkol, ranger, DNR Minong
Tom Michalek, fire chief, Wascott Volunteer Fire Department

Dean Nelsey, retired meteorologist, gathered weather data for 1977
Joni Orth, dispatcher, DNR Cumberland
Andy Podratz, tractor-plow operator, DNR Minong
Jim Rakitnichan, tractor-plow operator, DNR Minong
Larry Severtson, SEAT manager, Solon Springs Airport
Pat Sheller, forester, manager land and timber, Wausau Paper, Solon Springs
Dutch Snyder, pilot, SEAT turboprop airplane, Solon Springs Airport

People Interviewed Who Fought the Five Mile Tower

John DeLaMater, DNR Hayward area ranger—retired—fire boss on the Five Mile Fire
Bill Scott, DNR Minong ranger—retired—H Division boss on the Five Mile Fire
Ed Forrester, DNR Webster ranger—retired—"leapfrogging" ranger as boss of Divisions B, F, I, and K
Barry Stanek, DNR Gordon ranger—retired—boss of Divisions D, I, J, and L
Greg St. Onge, DNR Brule ranger—retired—boss of Divisions A and C
Ralph Mortier, forest-management staff specialist—retired (now deceased)—intelligence officer
Gene Miller, DNR Barnes ranger—retired—boss of Division C
Tom Quilty, DNR Hayward ranger—retired—boss of Division E
John Semo, DNR Pattison Park ranger—retired—boss of Division M
John Pohlman, DNR Ladysmith ranger—retired—boss of Division G
Barb Raasch, Spooner dispatcher—retired—was "Spooner Dispatch" for Five Mile Fire
Chuck Adams, DNR forester for Sawyer County—retired—plans boss
Dave Ives, DNR area fisheries manager, Hayward—retired—service boss

John Borkenhagen, DNR manager state tree nursery, Hayward—retired—supply officer
Gordon Landphier, chief of forest protection, Madison command center during fire
Duane Dupor, staff specialist, Bureau of Forestry, Madison command center during fire
Bob Becker, assistant district director DNR, Spooner office during fire
Don Jerabek, district staff specialist DNR, Spooner office during fire
D. J. Aderman, Hayward High school fire crew—now director of forestry for Johnson Timber, Hayward
Steve Balsick, Northwood High School fire crew
Lee Block, Northwood High School principal—retired
Jim Block, Northwood High School (Minong, Wascott, Gordon), mathematics teacher—retired
Jim Borkenhagen, Hayward High School fire crew
John Borkenhagen, Hayward High School fire crew
Bonnie Burnosky, Northwood High School fire crew
Les Chandler, Northwood High School fire crew
John Danielson, Hayward High School fire crew
Ralph Christiansen, DNR conservation warden—retired
Steve Coffin, forester, Mosinee Paper Industrial Forests—Solon Springs office—now DNR Bayfield County team leader
Don Crotteau, DNR Webster dozer-plow operator—retired
Joe Davidowski, DNR conservation warden—retired
Milt Dieckman, DNR conservation warden—retired
James Dienstl, DNR chief pilot—retired—airplane patrol in first hours of fire
Bud Erickson, DNR pilot—retired (now deceased)—airplane patrol on fire
Tim Foley, Northwood High School English teacher, later principal—retired
Everett Hanson, DNR Grantsburg dozer-plow operator—retired
Max Harter, DNR conservation warden—retired
Peg Herdt, Northwood High School fire crew
Bill Hoyt, DNR conservation warden—retired

Stan Johannes, DNR fish manager for Burnett and Washburn Counties, Spooner
John Kiel, DNR Gordon dozer-plow operator—retired
David Klaas, Northwood High School biology teacher
Ed Kofal, Northwood High School industrial arts teacher and guidance counselor—retired
Ron Kofal, DNR Gordon dozer-plow operator—retired
Don Monson, DNR North Spooner ranger—retired—boss of Division B
Ed Nelson, DNR conservation warden—retired
Pete Paske, DNR Minong dozer-plow operator—retired
Blaine Peterson, Northwood High School fire crew—now forester, Marathon County Parks and City of Wausau
George Phillips, DNR conservation warden—retired
Mark Radzak, Northwood High School fire crew—now forester
James J. "Bud" Schaefer, Northwood High School fire crew—now equipment technician, Minong DNR Ranger Station
John D. Schultz, camper who lit the match and accidentally started the Five Mile Tower Fire
Mark Smith, Grantsburg High School business education teacher—retired
Jim Stordahl, DNR assistant area forester for Douglas County
Bill Thornley, Spooner High School fire crew—now editor, *Spooner Advocate*
Kathy (Carr) Walker, Northwood High School fire crew—now owner, Walker Lumber, Minong
Allen Zaloudek, DNR North Spooner dozer-plow operator—retired

Author's note: I regret that I was unable to interview several people whose work on the fire was crucial but who were deceased by the time of the research for this book. Tom Roberts, DNR Brule area ranger, was the line boss on the fire. Conservation warden Tom White of Gordon passed away in 2007. My longtime friend George Becherer was a DNR Minong dozer-plow operator; others recall his skill and bravery. Bobby Hoyt, DNR Gordon dozer-plow operator, is mentioned in the book for his important contributions on the fire.

Other People Interviewed Who Fought or Witnessed the Fire

David Babcock, fought the fire
Eugene Barrett, witnessed fire and assisted evacuations of many people
Bernie Bergman, Wascott town chairman, firefighter
Kelly Blegen, watched fire as young girl in path of the flames
Harold "Buck" Block, business owner in Minong, fought the fire in dangerous conditions
Tom Brisky, fought the fire, owner of home in path of flames
Mark Cline, fought the fire
Michael Edwards, witnessed the fire
Kathy Fraatz, owner of Fraatz Busing and Trucking, organized sandwich making in Minong
Dottie and Bob Fries, owners of View Point Lodge at time of fire
Alex Grymala, fought the fire next to his lake
Emmy Herdt, school bus driver for workers during weekend and for students during the following three days
Jim Hill, Solon Springs Fire Department
Mary Hill, organized sandwich making in Solon Springs
Mike Johanson, deputy sheriff with Douglas County, assisted with evacuations
Dale Larson, witnessed the fire
Jean Legg, president of Minong Fire Department Auxiliary during the fire, Northwood School secretary
Jack Link, owner of Link's Grocery store, which provided food
Herbert Love, fought the fire with Douglas County Forestry Department
Joan Love, Gordon School custodian, witnessed the fire
Einere Marshall, school bus driver, witnessed the fire running past her home and property
Karen Matthias, wife of author, evacuated during the fire
Mark Matthias, son of author, witnessed the fire as a young boy
Jan Mednick, evacuated during the fire
Jim Mertes, witnessed the fire
Jack Morrow, witness the fire; along with his father, provided fuel
Mary Morrow, evacuated during the fire

Pete Pierce, owner of Loon Lake property that split the fire in two parts
Wayne Schirmer, fought the fire
Dick Scott, worked for Wolff Link at Link's store and delivered food to the firefighters
Kathy Scott, wife of H Division boss, organized sandwich making in Minong
Marilyn Scott, evacuated during the fire
Loren Sloan, fought the fire
Cheryl Smith, Northwood School secretary
Craig Smith, fought the fire with Gordon Fire Department crews
Harold "Smokey" Smith, fire department boss, Minong Fire Department
Mike Waggoner, crew leader on the fire
June Warner, Gordon School lead teacher, witnessed the fire
Mert Warner, Gordon Fire Department
Carol Wester, organized sandwich making in Solon Springs
Phil Wester, witnessed the fire
George "Hoppy" Wetzel, Minong Fire Department
Rod Wilcox, fought the fire
Nona Yrjanainen, owner of Nona's Café in Minong, provided food and coffee to firefighters

Appendix I

Five Mile Tower Fire Dispatch Summary

The following dispatch summary, copied word for word from handwritten notes in the DNR archives in Cumberland, Wisconsin, captures the first hours of the Five Mile Tower Fire, beginning with the first call, at 1:20 p.m. on April 30, 1977. The notes summarize the high-band radio communication between dispatcher Barb Raasch at the Spooner DNR dispatch center and fire boss John DeLaMater on his truck radio at fire headquarters.

1:20 p.m.—Call on emergency phone by Bill Scott
1:20—Report from Five Mile Tower of smoke
1:20—Tractor plow—stand by at Webb Lake
1:26—John Deere 450 tractor plow at Minong called
1:30—Tractor plow from Webster and one from Spooner called
1:31—North Spooner Ranger requested
1:31—Webster second tractor plow called
1:32—Webster ranger requested
1:32—Second Spooner tractor plow called
1:34—Ordered Hayward D-7 cat, Grantsburg 450 and Washburn Co. 450
1:35—Third Spooner tractor plow plus second Washburn Co. 450 t-p called
1:36—Ordered overhead team
1:39—Start ranger from Gordon
1:41—Grantsburg 450 t-p called
1:41—Bill Scott communication

1:42—Gordon Ranger Stanek communication
1:43—Minong tractor plow unit working on fire
1:47—Hayward heavy unit called, requested headquarters van, notify district office that project fire is in progress.
1:50—Gordon tractor plows called
1:58—Units arriving on fire
2:03—All emergency fire wardens in area notified to get hand crews called
2:03—Warden Flannigan called and offered to help—told to go to fire
2:08—Requested traffic control—called State Patrol
2:08—One heavy unit had tire blow out on Hwy 77
2:09—Webster units on fire scene
2:13—Requested 40 fire-fighters
2:20—Both Spooner tractor-plow units on fire scene
2:21—South Spooner ranger called
2:22—Forest Management bureau called to bring pick up load of back cans and shovels
2:31—Asked for Burnett County 450 t-p unit
3:06—Hayward station called
3:12—Hayward station requested Black River for return of Hayward area equipment
3:24—Asked for more traffic control—called Washburn and Burnett Co. Sheriff's Dept's.
3:25—Requested back cans and shovels from Hayward
3:30—Fire now 650 to 700 acres in size—requested 3 more t-p units plus 2 rangers
3:45—Mr. Becker coming from Spooner
3:45—Two t-p units and rangers coming from Brule area
4:22—Requested Gordon Prison Camp crew
4:39—Ladysmith t-p unit called
5:20—Additional tractor-plow units arriving on fire
5:30—Get news coverage for volunteers—someone to take pictures
5:32—D-7 cat from Penta Wood Products in Siren called
5:39—Get 100 more people
5:43—DNR Fish and Game crew on thc way up
6:09—Fire crossed county line (Washburn to Douglas)—traveled seven miles, fire is now three miles wide.

Time elapsed: 4 hours and 49 minutes: Rate = one mile burned every 41 minutes
7:14—Fire size reported at 4,800 acres after six hours
7:19—Two more heavy units report arriving on fire
8:50—U.S Forest Service crew from Trego arrived on fire
8:50—Get diesel fuel trucks

Appendix II

List of Those Who Fought the Five Mile Tower Fire

The following pages list 1,610 brave people who fought the Five Mile Tower Fire, from teenage fire crew members to senior citizens, from DNR personnel to local law enforcement, from trained firefighters to English teachers. Looking at the names, it is interesting to note that many family groups went to the fire to help on that devastating day: brothers, cousins, parents, and children.

I assembled the list from typed payroll lists, pencil notes, invoices, handwritten memos, and other documents in the DNR archives in Cumberland, Wisconsin. The numbers of hours worked come from DNR payroll sheets (hours worked by DNR employees are not included). Most of the volunteers were paid $2.30 per hour. For some, no payroll hours are listed in the records.

Approximately four hundred volunteer fire department personnel, fuel truck drivers, bus drivers, water tanker drivers, private dozer operators, National Guard dozer personnel, sandwich makers, and firefighters went directly to locations in the fire without signing up at fire headquarters, and unfortunately their names are not listed here.

I have taken great care to check spellings, and I apologize for any misspelled names, omissions, or mistakes that remain. If you're aware of any errors in this list, please e-mail corrections to billmatthias@hotmail.com for revision in a future edition.

On those dark days at the end of April and beginning of May 1977, thousands dropped what they were doing and volunteered to go into the black smoke and flames. This list is dedicated to the brave citizens of northwest Wisconsin who followed the smoke column, signed up, and went to work.

Name on DNR Payroll	Hours Worked
Abbott, Scott	12.0
Ackerson, Roger	31.0
Adams, Shirley	7.5
Aderman, Don	44.5
Adrian, Mike	3.5
Allen, Jack	27.0
Amundson, John	12.0
Amundson, Richard	15.0
Amy, Don	6.5
Anderson, A.R.	12.0
Anderson, Bruce	2.5
Anderson, Chris	8.0
Anderson, Elmer	9.5
Anderson, Gene	9.0
Anderson, Glen	8.0
Anderson, Harry	12.0
Anderson, Kenneth	15.0
Anderson, Loretta	11.5
Anderson, Lorne	12.0
Anderson, Mel	3.5
Anderson, Mike	46.5
Anderson, Norm	16.0
Andrea, Bob	17.5
Andrea, Sam	10.0
Apfel, Wesley	10.0
Arendt, Douglas	15.0
Armstrong, David	12.0
Armstrong, Kelly	13.5
Armstrong, Steve	9.0
Ash, Ted	10.0
Asp, Leslie	8.0
Asp, Martin	8.0
Asp, Ted	8.0
Atkinson, Vicki	12.0
Aubart, Charles	12.0
Aubart, Dave	21.0
Aubart, John	23.0
Aubert, Dennis	12.0
Augustyniak, Ray, Jr.	18.5
Avery, Paul	12.0
Avey, Ella	14.0
Ayers, Terry	18.0
Babcock, Dave	14.5
Bailey, Bill	12.0
Baker, Mike	29.0
Balcsik, Kim	42.5
Balcsik, Steve	61.5
Barbee, Fred	25.5
Barbee, Scott	25.5
Barbee, Tony	30.0
Barber, Jim	14.5
Bardine, Mark	12.0
Barkey, Warren	12.0
Barnes, Craig	5.5
Barret, Byron	12.5
Barrett, Dave	65.0
Barrett, Don	12.0
Barrett, Jerry	4.5
Barrett, Kay	9.0
Barrett, Marvin	21.0
Barsamion, Larry	3.0
Barshack, James	5.0
Bartel, Loren	5.0
Bartels, Harvey	12.0
Barthel, Kurt	95.5
Barthel, Tom	2.0
Bartle, Colleen	8.0
Bastian, John	7.5
Basty, Dave	6.0
Basty, John	6.0
Bates, Dennis	12.0
Bauch, Dave	10.0
Bauch, Tom	10.0
Bauer, John	18.5
Bauer, Ron	13.0
Baumgarten, Richard	5.0
Bear, Kurt	3.5
Bear, Phil	3.5
Beaupre, Scott	10.0
Becherer, Lori	27.0
Becherer, Loy	35.5
Beckett, Kelly	29.5
Beckett, Robin	8.0
Beeson, Bob	14.0
Belille, Frank	12.0
Belille, George	26.0
Belshaw, Bob	17.0
Belshaw, Cindy	11.5
Benjamin, Tim	11.5
Benjamin, Tom	11.5
Bennitt, William	14.5
Benrud, Bruce	8.0
Benson, Bruce	7.0
Bergmann, Axel	11.0
Bernston, Brian	22.5
Berquist, Dean	51.0
Berthiaume, Jeff	12.0
Biros, Dale	5.5
Biter, Karen	12.5
Bitney, Mike	15.5
Bixby, David	19.5
Black, Rick	67.5
Blanchette, Dale	8.5
Blaylock, Larry	8.0

Name on DNR Payroll	Hours Worked
Blaylock, Richard	8.5
Block, Harold	12.0
Block, James	59.0
Block, Lee	28.0
Block, Micki	12.5.0
Block, Ron	7.0
Boehm, John	14.5
Bohmert, John	3.5
Boland, Rick	15.5
Borkenhagen, Jim	20.0
Borkenhagen, John	20.0
Bos, Don	15.0
Bos, Ronald James	18.5
Bouchard, Jerome	12.0
Boughner, Jim	18.0
Bown, Rose	28.0
Boyd, John, Jr.	8.0
Boyd, John, Sr.	8.0
Brackin, Alfred	26.0
Bradmann, Randy	25.5
Brandine, Scott	12.0
Brandl, Charmaine	19.0
Brandl, Linda	9.5
Brandl, Randy	34.5
Brandt, William	6.0
Brekhe, Richard	18.5
Bresina, Rick	13.0
Bressler, Carol	25.0
Bridges, Bryan	32.0
Briggs, Byron	19.0
Briggs, Larry	19.0
Brigner, Don	28.0
Brisky, Julie	27.0
Brisky, Tom	12.0
Broadfoot, Tami	12.0
Bronson, David	17.0
Brown, Charlie	12.0
Brown, Dave	9.0
Brown, Donna	22.5
Brown, Harold	12.0
Brown, Hurley	9.5
Brown, Lorn	5.0
Brown, Ralph	20.0
Brown, Richard	13.0
Brown, Robert	2.0
Brown, Steve	13.5
Bruce, Don	11.5
Brueggeman, Mike	5.5
Brunes, Dave	7.0
Bruss, Terry	3.0
Buchman, David	14.0
Buchman, Dell	11.5
Buchman, John	12.0
Buck, Joe	19.5
Bull, Mike	8.0
Burfield, Leonard	31.5
Burling, Brad	2.0
Burnosky, Bonnie	25.0
Burnosky, Denise	16.5
Buros, Greg	10.0
Burr, Ken	3.0
Busch, Steve	12.0
Butler, Gary	8.0
Butterfield, Charlotte	23.5
Butterfield, Cheryl	12.5
Butterfield, Clara	23.5
Butterfield, Dewayne	11.5
Butterfield, La Roy	7.0
Byberg, Loren	14.0
Byrd, Steve	2.0
Cable, Craig	6.5
Cable, Joel	10.0
Cable, Peggy	19.5
Cable, Penny	19.5
Cairns, Dwayne	9.0
Campbell, Dave	9.0
Canfield, Mark	18.5
Cariolano, Jim	22.5
Carlaw, Laura	11.0
Carls, Linda	16.5
Carlson, Arne	1.5
Carlson, Brian	1.5
Carlson, Dale	1.5
Carlson, Judy	1.5
Carr, Jayne	12.5
Carr, Kathy	22.0
Carson, Gary	7.5
Carson, Gayle	7.5
Carson, Gayle, Jr.	7.5
Carson, Mike	17.0
Carter, Dick	12.0
Cartwright, John	12.0
Casler, James	12.0
Cassler, Mark	4.0
Catt, Duane	12.0
Chaffee, Mike	4.0
Chandler, Beth	12.5
Chandler, Jim	23.5
Chandler, Les	52.5
Chandler, Tom	62.5
Chenoweth, Rick	12.0
Cheple, Mark	4.5
Chide, Tom	12.0
Christiansen, Leonard	12.0

Name on DNR Payroll	Hours Worked
Christianson, Kim	29.0
Christner, Rebecca	14.0
Christner, Winton	27.5
Christoferson, Odell	38.0
Clark, Rusty	49.0
Clark, Scott	48.5
Clay, Leonard	12.0
Clayton, Wesley	85.0
Clements, Sam	20.5
Clements, Willis	12.0
Cline, Mark	4.5
Cloves, Mel	7.0
Cody, Norm	49.5
Colby, Larry	16.0
Collins, Ron	12.0
Collins, Scott	18.5
Combs, Roy	64.0
Comer, Tim	14.0
Conaway, Lon	4.5
Constance, Mark	12.0
Cook, James Jay	12.0
Cook, Ralph	12.0
Cosgrove, Kelly	10.0
Coughlin, Pat	10.0
Courrier, Fred	12.0
Cramey, Beth	23.5
Cramey, Bob	15.0
Cramey, George	5.5
Cramey, Mike	16.5
Crandell, Randy	28.0
Crowder, Robert	5.5
Crum, Wesley	8.0
Cunningham, Frank	12.0
Curtis, Al	3.5
Curtis, Albert	9.5
Cusick, Alan	5.0
Cusick, Tim	10.5
Dahlberg, Richard	7.0
Dahlstrom, Judy	11.5
Dahnert, Stan	4.0
Dale, Paul	12.0
Dalsveen, Edward	8.0
Daniels, Brian	18.5
Daniels, Chris	28.0
Daniels, Doren	8.0
Daniels, Julie	21.0
Daniels, Randy	7.5
Daniels, Rick	30.0
Daniels, Steve	6.5
Daniels, Sue	22.0
Danielson, John	13.5
Dansereau, Barb	50.5
Dansereau, Edmond	55.0
Dansereau, Elie	11.5
Dansereau, Renee	33.5
Dansereau, Rollo O.	21.0
Darby, Jim	22.5
Dasler, Ted	37.0
Davis, Jerry	18.5
Davis, Paul	16.0
Day, Frank	23.5
Day, Richard A.	12.0
Day, Victor	17.0
Degerman, John	10.0
DeMar, Linda	24.5
Demars, Anthony	17.5
Dempsey, Chris	11.5
Denham, Tom	31.0
Denkmann, Steven	8.5
Denniger, Clint	12.0
Denninger, Doug	31.5
Denninger, Vicki	15.0
Dennis, Arnold	12.0
Dent, Carolyn	10.0
Dent, Sherri	25.0
Dent, Sue	22.5
Dent, William L.	19.0
Derrick, Gary	19.5
Derrick, Kevin	9.0
DeSmith, Robert	8.5
Destache, Doug	1.0
Dezek, Bill	21.0
Dezek, Ray	11.0
Dhein, Fred	12.0
Dillie, Ernie	8.5
Dinnier, John	10.5
Doehr, Roger	10.5
Doehr, Vern	8.5
Dominowski, Dominic	10.0
Dominowski, John	28.0
Donattell, Allan	12.0
Donlin, John	12.0
Donnlyn, John	12.0
Dope, Charles	12.0
Doriott, Albert	4.0
Doriott, DeLroy	32.0
Doriott, Rick	8.0
Dorsey, Ken	12.0
Dorum, Larry	12.0
Dorum, Robert	12.0
Doskey, Joe	10.0
Draves, Donald	14.5
Dryberg, Doug	12.0
Dube, Greg	11.5

Name on DNR Payroll	Hours Worked
Duch, Terry	11.0
Duffy, Pat	7.0
Durand, Gary	10.5
Durand, Mike	13.0
Durand, Richard	3.5
Durham, Bob	3.5
Durham, Larry	3.5
Dziennik, John	14.0
Earl, Milton	12.0
Edison, Don	23.0
Eischen, Mark	7.5
Elliott, Dan	11.5
Elliott, Mike	12.0
Elliott, Tom	11.5
Ellison, Jeff	9.5
Elmlinger, Charles	7.0
Elmore, Jay	16.0
Elmore, Jim	29.5
Elmore, Jon	29.5
Elza, Steven	8.5
Enders, Ralph	4.5
Engel, David	4.5
Engel, John	4.5
Engel, Larry	24.0
Engles, Elmer	16.5
Engstrom, Mike	4.5
Ennis, Bill	18.0
Ennis, Mary	29.5
Ennis, Pat	26.5
Ennis, Susan	18.0
Erazmus, Eddie	22.0
Erickson, Mary	8.5
Erickson, Ron	8.5
Erskin, Walter	14.5
Everett, Chuck	12.0
Eytheson, Al	12.0
Eytheson, J.H.	12.0
Eytheson, Walter	12.0
Fairfield, Bridget	10.0
Fairfield, Eric	10.0
Farb, Eldon	6.0
Farrell, John	4.5
Faul, Eugene	10.0
Featherly, Alice	21.0
Featherly, Lori	29.0
Featherly, Phil	5.5
Featherly, Tammy	25.0
Fehr, Dave	12.0
Ferguson, Darrin	24.5
Ferguson, Diane	27.5
Ferguson, Dolean	27.5
Feuerhake, Dean	5.0

Name on DNR Payroll	Hours Worked
Fink, Mike	10.0
Finstad, Betty	11.5
Finstad, Bruce	15.5
Flamang, Herbert	12.0
Flavell, George	8.5
Flodin, Dan	12.0
Flynn, Pat	8.5
Foley, Timothy	31.0
Follis, Tim	12.0
Ford, Jess	3.5
Ford, Robert	3.5
Fornengo, A.J.	10.5
Fornengo, Martin A.	12.0
Fornengo, Marty	11.5
Foster, Georgia	16.0
Fox, Arland	18.5
Fox, C.W.	20.0
Fox, Gregg	17.5
Fox, Mary	7.5
Fraatz, Gerald	29.0
Freer, Randy	7.0
Frey, Henry	12.0
Frey, Tim	12.0
Friedman, Beverly	5.5
Frier, Randy	7.0
Friermood, Greg	11.0
Fries, Bob	13.0
Frigen, Dave	18.5
Fritz, Bill	3.0
Froemel, Larry	6.0
Froemel, Roger	1.0
Frost, Dale	10.0
Frost, Ron	10.0
Fry, Robert	14.5
Frye, Gus	5.5
Frye, Roger	12.5
Frye, Tammy	21.5
Frye, Tom	32.5
Fuller, Allen	8.0
Fye, Bill	25.0
Fye, Lori	17.0
Gablauf, Rick	12.0
Gabler, Rick	12.0
Gall, Larry	15.0
Gall, Tom	22.0
Garber, Dan	8.5
Garvey, Scott	15.0
Gates, Kay	12.0
Gatlin, Brenda	9.5
Gatlin, Homer, III	18.0
Gatlin, Homer, Jr.	18.0
Gatlin, Jim	17.5

Name on DNR Payroll	Hours Worked
Gatlin, Jim, Jr.	17.5
Gatlin, John	17.5
Gaulke, Robert	11.5
Gehrke, Lee	5.5
Gerberding, Dick	8.0
Germann, Shawn	4.5
Geving, Ray	2.0
Gilberg, Dan	18.5
Gilbert, Tod	15.0
Gillette, Jack	10.0
Glienke, Dave	13.5
Glienke, Dennis	13.5
Glienke, Doug	13.5
Goddyn, Elmer	1.0
Goddyn, Jim	1.0
Goldsmith, Larry, Jr.	12.0
Gomulak, Walter	12.0
Gorski, Jim	12.0
Gougar, Walter	11.0
Grafton, Duane	15.5
Gratson, Michael	6.0
Graves, Roy	3.0
Gray, Earl	12.0
Greely, Scott	20.5
Green, Dave	9.0
Green, Dave	12.0
Green, Greg	23.0
Greenhow, Ron	15.5
Griffin, Troy	11.5
Groat, Debby	24.0
Groat, Lyle	18.0
Groat, Lyman	15.5
Groat, Richard	32.5
Groth, Lyle	6.0
Grube, Ed	3.0
Grube, Pat	17.5
Gruebele, Clyde	15.0
Gucinski, Mike	6.5
Guildenzoph, Rollin	12.0
Gulseth, Fred	36.5
Gunderson, Ralph	34.0
Gutowski, Dianne	12.0
Gutowski, Tony	12.0
Guyette, Helen	24.0
Hadlock, Al	18.5
Haime, Bill	27.0
Haime, Walt	4.0
Hallman, John	4.0
Halonie, Dale	9.0
Hanacek, Cully	27.5
Hand, Steve	7.5
Hankins, Matthew	22.5
Hansen, Bill	5.0
Hansen, Ken	28.5
Hansen, Ogren	24.0
Hanson, Jerold	26.0
Hanson, Karen	24.0
Hanson, Mark	15.0
Hanson, Mary	12.0
Hanson, Sherri	10.0
Hanson, Susan	28.5
Hart, Fred	10.5
Harvey, Don	12.0
Harwood, Ron	12.0
Haseltine, Belva	7.5
Haseltine, Tom	7.5
Hatfield, Kurt	1.0
Haugen, Marvin	12.0
Hause, Robert A.	12.0
Hawkins, James M.	12.0
Hayes, Mike	6.5
Hazuka, John	10.0
Heath, Scott	2.0
Hebert, Bill	15.5
Hedrick, Jack	12.0
Hegg, Russel	7.5
Hein, Jim	12.0
Heineman, Stuart	20.5
Heinz, Herman	5.5
Heller, Oswald	29.0
Hemning, Dick	12.0
Hendrichs, Bob	11.5
Hendricks, John	5.5
Hendricks, Kenny	7.0
Hendricks, Lloyd	6.0
Hendricks, Richard	5.5
Hendrickson, Mike	7.0
Hendry, John	8.0
Hendry, Zim	8.0
Henke, Joan	6.0
Henson, Bruce	5.0
Henson, Frank	31.0
Henson, Linda	40.0
Henson, Tom	10.0
Herbert, Bill	11.5
Herdt, David	25.0
Herdt, Peggy	31.5
Herdt, Sharon	26.5
Herman, Glen	10.5
Herman, Tracy	33.0
Higgins, Gordon	14.5
Hill, David	15.0
Hills, Harold	24.5
Hills, Howard	14.5

Name on DNR Payroll	Hours Worked
Hills, Mike	14.5
Hinde, George	15.5
Hoag, Doug	10.0
Hoag, Russel	12.0
Hoelter, Gene	8.5
Hoerning, Frank	7.5
Hoffmann, Greg	12.0
Holcomb, Basil	14.5
Holcomb, LeRoy	7.0
Holly, Louis	8.0
Holt, Karla	29.0
Hopkins, Everett	22.5
Hort, Lisa	15.5
Howard, Greg	10.0
Howe, Kent	27.0
Hoyt, Patrick	12.0
Hruska, Jim	12.0
Hubbell, Rick	10.5
Hubbell, Roy	12.0
Hudek, John	7.5
Huffer, Wes	19.0
Huftel, Gary	12.0
Huftel, Terry	12.0
Hughes, Cliff	6.5
Hughes, Jim	8.0
Hunter, Stu	12.0
Huntowski, Peter	31.0
Huntowski, Tom	45.0
Huray, Dan	10.0
Huray, Leonard	10.0
Irvine, Lori	25.5
Irvine, Lorie	11.5
Irwin, Morris	12.0
Jackson, Phillip	8.5
Jackson, Roy	4.5
Jackson, Ward	12.0
Jacobs, Donald	6.0
Janke, Sandy	22.5
Jarosch, Ted	12.0
Jarvis, Leon	12.0
Jensen, Bill	12.0
Jensen, Mike	15.0
Jerome, David	7.0
Jewell, Bob	9.0
Joadwine, Tim	25.0
Johnsen, Bill	15.0
Johnsen, Brian	24.5
Johnson, Aldon	12.0
Johnson, Bill	12.0
Johnson, Brian	12.0
Johnson, Bud	12.0
Johnson, Dave	11.0
Johnson, Dean	12.0
Johnson, Denise	12.5
Johnson, Jack	22.0
Johnson, Keith	9.0
Johnson, Lee	4.0
Johnson, Lester	11.0
Johnson, Lori	23.5
Johnson, Mark	8.5
Johnson, Mike	6.5
Johnson, Paul	28.0
Johnson, Randy	25.5
Johnson, Scott	20.5
Johnson, Ted, Jr.	18.0
Johnson, Tim	9.0
Johnson, Vernon	5.0
Johnson, William M.	12.5
Jonas, Paul	7.0
Jones, Bill	31.0
Jones, Gene	1.0
Jones, Kathy	2.0
Jorgensen, Dave	5.5
Josephs, Jeff	12.0
Jung, Don	6.0
Jung, Sally	15.5
Jung, Tim	6.0
Kagigebi, Charles	12.0
Kaiser, Pat	6.5
Kamin, Brian	76.5
Karow, Jim	16.0
Kartchmer, Al	8.5
Kartchmer, Marlin	8.5
Kasinskas, Marty	19.0
Kasten, Jeff	11.5
Katzmark, Cheryl	12.5
Katzmark, Scott	48.5
Kauffman, Lilah	9.0
Kegel, Ann	15.5
Keller, Kim	8.5
Keller, Ron	3.5
Kelley, Robert	12.0
Kelling, Craig	8.0
Kelling, Earl	8.0
Kelsey, Holly	22.5
Kelsey, Mark	29.5
Kemp, Tom	12.0
Kersten, Rod	25.0
Kerzich, Mark	12.0
Kessler, Darlene	11.5
Kessler, Tim	25.5
Kevan, Cathy	23.5
Kevan, Neil	12.0
Kevan, Tim	45.0

Name on DNR Payroll	Hours Worked
Kibble, Bob	12.0
Kibble, Carol	12.0
Kibble, Edna	12.0
King, Katy	11.5
King, Lee	56.0
King, Marvin	2.0
King, Millie	52.5
Kinnear, John	10.0
Kinney, Valorie	31.0
Kisselburg, Keith	31.0
Klaas, David	27.0
Klawitter, Erwin	8.0
Klawitter, Gene	11.0
Klawitter, Ray	12.0
Kline, Frank	10.0
Kline, Jeff	61.0
Klippel, Paul	13.5
Klobertanz, Robert	19.0
Klugow, B.F.	15.0
Klugow, Bill	15.0
Knick, Jack	6.5
Knight, Robert	5.0
Knoop, Glen	17.5
Knutson, Bill	14.0
Knutson, Mark	14.0
Knutson, Stan	6.0
Koedyker, Fran	24.0
Koedyker, Ron	24.0
Koel, Richard	14.0
Kofal, Bunk	21.0
Kofal, Ed	28.0
Komurka, Van	5.5
Konczal, Andy	12.0
Kopecky, Howard	12.0
Kornfeind, Ray	12.0
Kornfeind, Tony	20.0
Kornfind, Tony	9.0
Korzenieski, Harry	12.0
Kosnak, Scott	28.5
Kramer, Leona	51.5
Kramp, John	9.0
Kranitz, Ron	4.5
Kranz, Jim	12.0
Kreb, Duane	7.0
Kreb, Ken	7.0
Krencisz, Sandy	27.0
Krenz, Craig	10.0
Krenz, Kim	10.0
Krisak, Gary M.	18.0
Krisak, John	6.5
Krisak, Roger	6.5
Kroeze, Chester	38.0
Krondlund, Mark	9.5
Krueger, Mike	12.0
Kruger, Jeff	11.5
Kruger, Robin	11.5
Krumviede, Lyle	32.5
Ksobiech, Roy	9.0
Kuba, Cindy	12.0
Kuba, Don	12.0
Kuba, Don, Jr.	35.0
Kuba, Jeff	12.0
Kubala, Jim	12.0
Kuczenski, Paul	8.5
Kunz, Blanche M.	23.5
Ladenthin, Robert	12.0
Lamberg, Kurt	12.5
Landwehr, Ron	15.0
Lange, Arthur	8.0
Lantz, Shirley	18.0
LaPorte, Jeffrey	6.5
LaPorte, Melvin	23.0
Larrabee, Jim	18.0
Larson, Gary	22.0
Larson, Gene	14.0
Larson, Jeff	4.0
Larson, Jeffrey	12.0
Larson, Jim	7.0
Larson, Leonard	12.0
Larson, Lynn	12.0
Larson, Randy	12.0
Larson, Tom	16.0
Larson, Walter	16.0
Latuff, Mike	9.5
Lawler, Wayne, Jr.	7.5
Lawler, Wayne, Sr.	7.5
Lawrence, Larry H.	30.0
Ledbeter, Lawrence	12.0
Lee, Hugh	8.0
Lee, Laura	14.5
Lee, Mike	18.5
Leech, Len	12.0
LeFebore, David	4.5
LeFebore, Mike	8.5
Legg, Bill	20.0
Lemon, Floyd	12.5
Lemon, Mike	32.0
LeMone, Dave	15.5
Lencowski, John	12.0
Lencowski, Mike	12.0
Lester, Stephani	15.0
Leusman, Judy	10.0
Levake, Larry	20.5
Lewis, Ben	5.5

Name on DNR Payroll	Hours Worked
Lieber, Dan	12.0
Lieber, Ralph	12.0
Linderfelser, Jeff	7.5
Lindquist, Maynard	15.0
Lindstrom, Chris	8.5
Little, Pat	3.0
Litwin, Dave	18.0
Litwin, Ed	16.0
Livingston, Kevin	9.0
Livingston, Mark	15.0
Lofkvist, Dawn	8.5
Lofkvist, Howard	8.5
Lord, John	12.0
Love, Ben	59.5
Love, Jeff	19.0
Love, Robert	31.0
Love, Vicky	30.0
Lozen, Gerald R.	6.5
Lucarelli, Pete	21.5
Luebbe, Donald	21.0
Lueck, Stanley	10.0
Luedtke, Dan	7.0
Lunderville, Dave	12.5
MacDonald, Laurie	4.5
Macho, Dick	12.0
Mack, Linda	21.5
Mack, Robert	25.0
Mahmke, Earl	9.0
Main, Connie	6.0
Main, Dennis	13.0
Main, Greg	15.0
Main, Larry	6.0
Maina, David	8.5
Maina, Richard	8.5
Maley, Eric	18.0
Manderschied, Harold	12.0
Marciulionis, Ray	8.0
Marcucheck, Teri	14.0
Marholz, Dan	4.5
Marion, Bill	12.0
Markgen, Phil	13.5
Markgen, Steve	8.5
Marquardt, William	17.5
Martell, Jim	9.0
Martin, Dave	3.0
Martin, Kevin	24.5
Martin, Richard	10.0
Martinson, Lee	22.5
Masten, Les	11.5
Matthias, Bill	58.0
Mattie, Collyn	48.0
Mattila, John	6.5
Maus, Dick	25.0
McCumber, Bruce	29.0
McCumber, Jim	49.0
McCumber, Mary	22.0
McCusker, Joe	34.0
McDowell, Mark	31.0
McDowell, Merritt	19.5
McDowell, Neva	58.0
McFarren, Guy	17.0
McFarren, Scott	6.0
McGraw, Jack	23.5
McIlguham, Mike	14.5
McKelvey, James	11.5
McLaughlin, Frank	16.0
McLaughlin, Terry	26.0
McQuade, Randy	8.5
McQuade, Sam	8.5
McQuade, Tim	25.5
McShane, Mike	24.0
Mednick, Charles	12.0
Mehsikoler, Debbie	18.0
Meier, James	21.0
Meitzner, Jane	10.0
Melhuse, Jon	11.0
Melton, Bradley	9.5
Melton, Chet	11.5
Melton, Clarence	13.0
Melton, Delray	9.5
Melton, Jay	6.0
Melton, Michael	9.0
Mendyke, Doug	11.5
Meronk, Dick	10.5
Metcalf, Curt	22.5
Metcalf, John	8.5
Michaud, Jack	12.0
Micken, Bill	19.5
Mier, Kurt	9.0
Milbert, Jeff	15.0
Milkert, Billy	7.5
Milkert, Bob	7.5
Miller, Cary	16.5
Miller, Greg	21.0
Miller, Jim	25.0
Miller, Joe	25.0
Miller, Kent	8.5
Miller, Rita	7.0
Miller, Steve	21.0
Mitchel, Sue	1.0
Moats, Don	34.5
Moen, Dave	12.0
Monnier, David	31.0
Monnier, Earl	18.0

Name on DNR Payroll	Hours Worked
Monnier, Hazel	28.0
Monnier, Ron	4.0
Monnier, Tim	27.0
Moore, Jerry A.	13.0
Moore, Randy	13.0
Morehouse, Robert	8.5
Morgan, Doug	8.5
Morris, Jesse	26.0
Morse, Ken	17.5
Morse, Wesley	26.0
Mortier, Jeff	22.0
Moser, Jerry	8.0
Moser, Pat	12.0
Moser, Pete, Jr.	8.0
Moyer, Allen	31.0
Moyer, Samuel	11.5
Mueller, Daune	12.0
Mueller, Mark	7.0
Mulleur, Robert J.	74.0
Murawski, Betty	27.0
Murawski, Jim	8.0
Murray, John	23.5
Musolf, Jeff	12.0
Myers, Dan	15.0
Mysicka, Joe	12.0
Mysicka, Tom	12.0
Nelson, Andy	25.0
Nelson, Arvid	12.0
Nelson, Cheryl	27.0
Nelson, Doug	15.0
Nelson, Mel	3.5
Nelson, Roy	12.0
Nelson, Sue	48.0
Nelson, Terry	12.0
Newman, Mike	12.0
Newson, Bill	20.0
Newton, Mike	9.5
Nichols, Charles	11.5
Nieckula, Jim	1.0
Nielson, Barry	5.5
Nielson, Steve	15.0
Niemuth, Kathy	12.0
Nikstad, Timothy	6.5
Nimmer, John	10.5
Nordstrom, Ed	11.0
Nordstrom, Ross	20.5
Norquest, Andrew W.	11.0
Norton, Kris	15.0
Norton, Terri	2.0
Nutt, Doug	11.0
Nutt, Stanley	11.0
Nutt, Wayne	11.0
Nyren, Don	12.0
Nyren, Ken	12.0
Oakes, Da	12.0
Ogren, Roger	81.5
Oimston, Bill	12.0
Okonek, Dave	12.0
Okonek, Don	8.0
Okonek, Jerry	8.0
Oliverson, Gil	12.0
Olsen, Dave	7.5
Olson, Bill	18.0
Olson, Gary	8.5
Olson, Larry	6.5
Olson, Merryn	8.0
Olson, Randy	4.5
Olson, Rick	32.5
Olson, Scott	12.0
Omernik, Jerome	11.0
Onczal, Andy	12.0
Organ, David	17.5
Organ, Theresa	12.0
Orlin, Bruce	7.0
Orlin, Suzan	7.0
Ounneron, Ralph	12.0
Oustigoff, Leva, Jr.	8.0
Oustigoff, Leva, Sr.	8.0
Overman, Russ	11.0
Owczynski, Andrey	12.0
Packard, Richard	7.5
Paffel, Tony	6.0
Pagorek, Tony	19.0
Pardum, Donald	12.0
Parise, Eugene	7.5
Parise, Pat	7.5
Parker, Howard K., Jr.	19.0
Partridge, Jim	7.0
Patrick, John M.	12.0
Pattee, Donald	17.5
Pattee, Robin	8.5
Pattee, Ron	8.5
Paul, Dale	14.0
Paulson, Harold	7.0
Paulson, Jodi	11.5
Pease, Bob	21.5
Pease, Toby	17.0
Peck, Jack	4.0
Peck, Jim	4.0
Peckham, Steve	8.5
Pederson, Howard	12.0
Pederson, Larry	27.0
Pederson, Roy	12.0
Peitz, Dennis	8.0

Name on DNR Payroll	Hours Worked
Pembrook, Rick	9.0
Pepin, Mike	4.5
Peppler, Gary	11.5
Perry, Dale	9.0
Perry, Richard	14.5
Pesko, Mike	6.0
Peterson, Blaine	17.0
Peterson, Charles	12.0
Peterson, Deb	12.0
Peterson, Gary	7.0
Peterson, Jerry	17.5
Peterson, Roger	1.5
Peterson, Shelly	12.5
Peterson, Tom	12.0
Peterson, Toni	14.5
Petrie, Lori	16.0
Petry, George	9.0
Pettingill, Bucky	13.0
Pfiester, Dick	12.0
Pfiester, Ronnie	12.0
Phenn, Allen	10.0
Phermetton, Lloyd	8.0
Philip, Bob	7.5
Philips, Bob	2.0
Pierce, Brian	28.0
Pies, Jim	20.0
Pleau, George R.	12.0
Plesums, Juris	11.5
Pluss, Charles	12.0
Polaski, Mike	7.0
Polaski, Pete	12.0
Pollard, William	12.0
Pollock, Mark	10.0
Pollock, Mike	10.0
Poppe, Cliff	12.0
Poppe, Fred	12.0
Poppe, Larry	12.0
Poppe, Michael	12.0
Porter, Ray	10.0
Poslusny, Tom	7.0
Post, John	7.0
Postl, Karen	24.0
Postl, Keith	39.0
Potzen, Carl	28.5
Poulin, Russel	8.5
Powers, Tim	16.5
Predni, Dick	8.0
Prellwitz, Dwain	12.5
Press, Cindy	17.0
Price, Bob	12.0
Priem, Scott	10.0
Priem, Terry	10.0
Pringle, Errol	18.0
Propst, Ken	16.0
Pulvermacher, Ted	24.5
Pydo, Walt	22.0
Quagon, Tom	33.0
Quinn, Lanny	1.0
Radke, Dale	15.5
Radloff, Barry	8.0
Radloff, Dave	8.0
Radzak, Mark	65.0
Rampier, Linda	10.0
Rand, Bob	9.0
Randall, Terry	1.0
Randt, Ralph	15.0
Randt, Ralph	3.5
Rassmussen, Dennis	8.0
Rassmussen, Steve	6.5
Rediger, Gary	3.0
Reimer, Dan	5.5
Reinolt, Anitta	21.0
Reinolt, Marvin	14.5
Ricci, Jim	1.0
Rice, David	10.0
Rice, Robert	25.0
Rich, Dawn	25.5
Rich, Kim	25.5
Rich, Ron	12.5
Richard, Julius	27.0
Richardson, Del	25.0
Richardson, Scott	28.5
Richmond, Glen	7.0
Richmond, Roy	18.5
Ridgeway, Paul	8.0
Riedinger, Schawn	9.5
Riedinger, Sean	12.0
Riemer, Andy	14.5
Rigby, Lee	3.0
Rigby, Roger	3.0
Ringlien, Brent	13.0
Rinnman, Monte	28.0
Ritzinger, Paul	4.0
Rivera, John	9.5
Rivera, Robert	14.0
Robarge, Don	12.0
Robarge, Randy	24.5
Robbins, Steve	12.0
Roberts, Larry D.	12.0
Robinson, Patrick	14.5
Robotka, Emil	6.0
Robotka, Jim	6.0
Rodgers, Greg	18.0
Roessel, Allen G.	12.5

Name on DNR Payroll	Hours Worked
Rogers, Don	12.0
Rogers, Ray	12.0
Rogus, John	12.0
Rohrman, Bernard	16.5
Romberg, John	8.0
Rondeau, Vernon	1.0
Ross, Guy	12.0
Ross, Jerry	9.0
Rother, Clarence	12.0
Rothermel, John	12.0
Rouse, Kathy	15.5
Rouse, Ken	12.0
Rowe, Bill	9.0
Rowe, Jim	12.0
Rowe, Kevin	10.0
Rowe, Randy	10.0
Rudi, Wayne	12.0
Rufus, Hudson	12.0
Rule, Jim	13.0
Rupp, Bill	17.0
Ryberg, Dough	12.0
Ryberg, Glen	8.0
Sager, Terry	7.0
Saline, Terry	7.0
Salzman, Henry	9.5
Salzman, Larry	16.5
Samuelson, Gerald	3.0
Samuelson, Leo	3.0
Sanders, Dave	6.5
Sandlin, Frank	12.0
Sandlin, Judith	12.0
Sandman, George	7.5
Sanford, Vernon	36.0
Sannwald, Mary	11.5
Sayles, Terry	12.0
Scanlon, Mark	38.0
Scanlon, Mat	14.0
Scelly, Robert J.	12.0
Schaaf, Glen	13.0
Schaaf, Marty	20.0
Schaaf, Mike	20.0
Schaefer, Brian	8.0
Schaefer, Bruce	8.0
Schaefer, "Bud" James, Jr.	81.5
Schafhauser, Bill	6.5
Schafhauser, Bill, Jr.	6.5
Schafhauser, Jim	7.5
Schafhauser, Michael	6.5
Schalzo, Tim	12.0
Scharein, Ray	12.0
Schatzlein, Janet	4.0
Scheel, Joe	17.0
Scherer, Phil	4.5
Schinler, Mark	12.0
Schirmer, Mike	22.0
Schirmer, Wayne	25.0
Schmitz, Doug	9.0
Schmitz, Robert	26.5
Schmitz, Roger	35.5
Schmitz, Ronnie	26.5
Schmoeckel, Jim	3.5
Schoolmeesters, Dave	15.5
Schowolter, Larry	2.5
Schroth, Jerry	26.0
Schuenemann, Rick	12.0
Schuette, George	7.5
Schuffel, John	8.5
Schult, Ken	2.0
Schultz, Brian	18.5
Schultz, James	15.0
Schultz, Jim	12.0
Schwartz, Dan	14.0
Schwedersky, Mark	11.5
Schwedersky, Mike	11.5
Scott, Kathryn	40.0
Scott, Patty	17.0
Scott, Theresa	27.0
Scottum, Bob	7.0
Scribner, Tim	15.0
Sears, Bill	8.0
Seckora, Rick	11.5
Seckora, Robert	9.0
Seifert, Cindy	27.0
Seifert, David	26.5
Seifert, Lisa	26.0
Selberg, Ron	7.0
Semanko, Dave	1.5
Severson, Ron	12.0
Shafstall, John	7.5
Shaver, Bill	6.0
Shervey, Tom	12.0
Shireman, Jan	11.5
Shook, Mike	12.0
Shores, Ernest, Jr.	10.0
Shower, Laverne	12.0
Siebens, Ed	17.5
Siebens, Rita	17.5
Siebert, Warren	14.5
Sienko, Joe	6.0
Simmons, Dave	15.5
Simmons, Gus	12.0
Simmons, Jim	10.5
Simmons, Lee	12.0
Simmons, Pete	12.0

Name on DNR Payroll	Hours Worked
Sirek, Gary	13.0
Sirek, Scott	13.0
Skabroud, Richard	7.0
Skau, Ronald	12.0
Skoglund, Paul	7.5
Slinker, Dean	10.5
Slinker, Gary	10.5
Slivensky, Dan	10.0
Sloan, Loren, Jr.	33.0
Sloan, Robert L.	22.5
Smith, Bob	27.5
Smith, Brad	10.0
Smith, Chuck	12.0
Smith, Craig	14.5
Smith, Dee Ann	28.5
Smith, Diana	9.0
Smith, Greg	3.5
Smith, Jeff	15.5
Smith, John	20.0
Smith, Ken "Dewey"	17.0
Smith, Kerry	15.0
Smith, Lance	3.5
Smith, Lyndon	10.0
Smith, Mark	20.0
Smith, Mike	24.0
Smith, Nig	29.0
Smithberg, Morris	1.5
Snell, Joe	32.0
Snider, Richard	12.0
Snippen, Dave	10.0
Snyder, Randy	15.0
Sorchuck, Mike	10.0
Soule, Mark	7.5
Spieler, Dan	38.5
Splinter, Jim	11.5
Sporcia, Tony	14.0
Stariha, Tim	41.0
Stariha, Tony	28.0
Stauffer, Brian	11.5
Stauffer, Dan	21.0
Stauffer, Steven P.	19.0
Steen, Richard	2.5
Steffen, Tom	50.0
Steltz, Art	17.5
Stensland, Roger	9.0
Stewart, Connie	11.5
Stewart, Curtis	5.0
Stewart, Tom	10.0
Stoeckel, Carl	12.0
Stoeckel, Randy	6.5
Stone, Dana	14.5
Stone, Gary	3.5
Stordahl, Jeff	10.0
Strenke, Ray	1.0
Strenke, Sue	19.0
Strock, Kerry	16.5
Stubbs, Pat	7.0
Stubfors, Art	14.5
Studeman, Marvin	9.0
Stun, Elliot	12.0
Suhsen, Leon	18.0
Sullivan, Pat	25.5
Sundeen, Steve	14.0
Swanberg, John	10.0
Swanson, Bob	12.0
Swanson, Craig	12.0
Swanson, Rollin	12.0
Swearingen, Mike	12.0
Swonger, Gerald	7.0
Sybers, Richard	20.0
Tainter, Art	12.0
Tainter, Bill	12.0
Talbert, Elmer	12.0
Talbert, William	12.0
Taylor, John	1.0
Tepoel, Deb	25.0
Terbilcock, Bill	9.5
Terbilcock, Lory	8.5
Terry, Forest	1.0
Terry, Mike	14.0
Thannul, Jim	28.0
Thomas, Bruce	16.0
Thompson, Chris	18.0
Thompson, Jack	6.0
Thompson, Mary	17.0
Thompson, Michael	12.0
Thompson, Pat	12.0
Thompson, Richard	24.5
Thompson, Stuart	12.0
Thompson, Tim	7.0
Thoreson, Ronald	12.0
Thornley, Bill	14.0
Tindal, Jerry	12.0
Tindal, Tom	12.0
Tinsley, Harry	12.0
Todd, Angie	8.0
Todd, Terri	8.0
Tolene, Dan	24.0
Tollner, Wayne	13.5
Tourville, John	6.5
Townsend, Carolyn	23.5
Townsend, Leo	11.0
Townsend, Randy	39.5
Treise, Ed	15.0

Name on DNR Payroll	Hours Worked
Tucker, Larry	7.0
Tucker, Randy	14.0
Turek, Terry	15.5
Tuverson, Ron	10.0
Uren, Terry	1.0
Vacik, Ed	3.0
Van Etten, Warren	1.5
Van Guilder, Jim	19.5
Visger, Al	18.5
Visger, Dennis	6.5
Visger, Elmer	31.0
Visger, Gene	12.5
Visger, Jeff	12.0
Visger, Rodney	30.0
Vraniak, Jerry	6.0
Wachewicz, Larry	14.0
Wachewicz, Tony, Jr.	14.0
Wachewicz, Tony, Sr.	17.0
Waddell, Steve	65.0
Wade, Gene	9.5
Wade, Larry	5.5
Waggoner, Bill	4.5
Waggoner, Jarvis	12.0
Waggoner, Jim	15.5
Waggoner, Mike	38.0
Waggoner, Steve	19.5
Wahlquist, Fred	12.0
Wahlstrom, Bruce	18.5
Wahlstrom, Pat	7.0
Waldorf, Clayton	5.5
Waldorf, Clayton, Jr.	17.5
Wanek, Bob	13.5
Washkuhn, Greg	11.5
Watts, Albert	16.5
Weaver, Clark	12.0
Weaver, Kevin	12.0
Weaver, Lorie	12.0
Webb, Richard	12.0
Weber, Roger	12.0
Wehmhoefer, Cindy	12.5
Wehmhoefer, Rick	19.5
Wehrman, Larry	12.0
Weingarten, Del	21.5
Weisburg, Robin	11.0
Wentzel, Henry	9.0
Wentzel, Marion	17.0
West, Matt	10.0
Westbrook, Gerald	6.5
Westcott, Tom	17.0
Westpahl, Terry	4.5
Whitcomb, Dave	12.0
White, Don	25.0
White, Kevin	15.5
Whiteside, Mike	16.0
Whyte, Robert	8.5
Wick, Greg	11.0
Wierschem, Steve	5.0
Wiggs, Lane	4.0
Wilcox, Charles	12.5
Wilcox, Gary	12.5
Wilcox, LaVonne	12.5
Wilcox, Merril	12.5
Wilcox, Richard	12.5
Wilcox, Rod	12.5
Wilde, Lloyd	15.5
William, Bruce	12.0
William, Roger	9.5
Williams, Bill	11.0
Williamson, Ross	8.0
Wilson, Rod	11.5
Wiltermuth, Rod	4.0
Winn, Ron	16.5
Wozniak, Gerald	12.0
Wright, Peter	12.0
Wuethrich, Fred	14.0
Yackel, Steve	1.0
Yeazle, Cheri	11.5
Young, Darwin	29.0
Young, Dennis	6.0
Zaloudek, David	8.0
Zaloudek, Ed	8.0
Zarn, Karen	12.0
Zbytovsky, David	12.0
Zeien, Clem G.	12.0
Zeien, Kevin	23.5
Zeien, Rocky	23.5
Zillner, Ed	17.0
Zimit, Carol	29.5
Zimmer, Jim	13.0
Zimmerman, Randy	9.5
Zitzka, Wayne	7.0
Zunker, Dennis	10.0
Zunker, Elaine	10.0

DNR Employees

Adams, Chuck
Backhaus, Carl
Becherer, George
Becker, Bob
Blinkwolt, Gary
Borkenhagen, John
Bugenhagen, Jon
Campbell, Adeline
Clarke, Art
Crotteau, Don
Debrin, Howard
DeLaMater, John
Dieckman, Milt
Dienstl, Jim
Dillon, Tom
Donatell, Jack
Dries, Bob
Dunn, John
Erickson, Bud
Ericson, Cully
Ferber, Ed
Flanigan, Jim
Follis, Buck
Forrester, Ed
Gingles, Earl
Gothbald, Robert
Hanson, Everett
Hanson, Lowell
Harter, Max
Housel, Bob
Hoyt, Bill
Hoyt, Bob
Ives, Dave
Jerabek, Don
Johannes, Stan
Johnson, Hugh
Johnson, Wes
Kiel, John
Kling, Dan
Kofal, Ron
Lang, Lyman
Libby, Harry
Long, Lyman
Love, Jerry
Ludtke, Orlie
Luedtke, Orland
Lund, Gary
Magnuson, Ken
Miller, Gene
Monson, Don
Moore, Sam
Mortensen, Dave
Mortier, Ralph
Nelson, Bud
Nelson, Ed
Nordstrom, Arnold
Olmstead, Arnie
Olson, Don
Olson, Larry
Panqullo, Sam
Paske, Lyle
Pickert, Jack
Pohlman, John
Quilty, Tom
Riemer, Larry
Ritchie, William
Roberts, Tom
Sandberg, Duane
Sande, Jack
Savage, Pat
Scott, Bill
Semo, John
Senske, Ken
Smith, Glen
Soderberg, Duane
Stanek, Barry
Stevenson, Don
St. Onge, Greg
Stordahl, Jim
Tech, Dave
Tesky, Tom
Vallem, Ray
Varro, Jim
White, Bob
White, Tom
Wick, Bob
Williams, Bruce
Wilson, Chuck
Wilson, Dave
Witt, Pete
Zaloudek, Al
Zosel, Chuck

Index

Page numbers in *italics* indicate photographs.

About the Author

Photo by Karen Matthias

Bill Matthias was born in Columbus, Wisconsin; grew up in Madison; and now lives on Bond Lake in Wascott. He began his studies at the University of Wisconsin–Madison as a preforestry major and then rekindled his interest in the pine, oak, and lake regions of northwest Wisconsin when he became the superintendent of Northwood School District in Minong in 1975. While he was superintendent, Matthias launched the teenage firefighting crews at Northwood High School and battled the Five Mile Tower Fire of 1977 for fifty hours.

He and his wife, Karen, raised their four children, Mark, Greg, Kristin, and Jeff, on Bond Lake. Today they spend their winters on the salt water in Englewood, Florida, and each spring return to the Wisconsin Northwoods, where Matthias is a charter member of the Wascott Volunteer Fire Department.

Praise for *Monster Fire at Minong*

"A well-balanced account of a past event and its impact upon the present."

—JAMES R. MILLER, retired, Wisconsin DNR Fire Control

"I was a senior at Spooner High School, just south of Minong, when the fire broke out. I became one of the teenagers who volunteered to help fight the fire, and it must have made a real impact, because all these years later it is still clear in my mind. We spent two days and nights with water tanks on our backs, walking through charred timber lands putting our small fires. We were coal black from head to foot. Every once in a while a stump would 'explode' and scare the heck out of us. We watched rabbits and deer wandering aimlessly as if lost. It was an alien world.

"The fire seemed to have a life of its own, and I recall wondering how a bunch of puny humans could possibly deal with such a monster. Bill Matthias has captured that feeling of awe and dread in his wonderful account of an historic time in northwestern Wisconsin history. This book is a masterwork of detail that needed to be written, and I doubt anybody could have recreated the mood of the time better.... If you were there it is like reliving the moment, and if you were not there this is a story that will capture you and not let you go until you've finished the final sentence."

—BILL THORNLEY, editor, *Spooner Advocate,*
and teenage firefighter on the Five Mile Tower Fire